HONDA/ACURA ENGINE PERFORMANCE

How to Modify D, B and H Series Honda/Acura Engines for Street and Drag Racing Performance

Mike Kojima

HPBOOKS

HPBooks
are published by
The Berkley Publishing Group
A division of Penguin Putnam Inc.
375 Hudson Street
New York, New York 10014

First edition: April 2002
ISBN: 1-55788-384-X

10 9 8 7 6 5 4 3

This book has been catalogued with the Library of Congress

Book design and production by Michael Lutfy
Cover design by Bird Studios

Interior photos by author unless otherwise noted

NOTICE: The information in this book is true and complete to the best of our knowledge. All recommendations on parts and procedures are made without any guarantees on the part of the author or the publisher. Tampering with, altering, modifying or removing any emissions-control device is a violation of federal law. Author and publisher disclaim all liability incurred in connection with the use of this information.

ACKNOWLEDGMENTS

(In memory of Matt Kempe 1979–2000)

I would like to thank my wife Sharon and daughter Christa for letting me take the time to write this book; Ken Nord at American Honda for correcting my Honda mistakes; Brian Gillespie of Hasport for helping me figure out the amazing amount of engine swap and JDM engine part combos; Rob Cadle at Garrett Engine Boosting Systems for teaching me the engineering of turbochargers and checking my math; Darren San Angelo of R&D for letting me take exclusive peeks inside his super-fast and secret NA motors; Paulus Lee and Ben Ma of AEBS for letting me snoop around their secret lab; and finally, the guys and gals that make it all happen, the racers: Lisa and Gary Kubo, Steff Papadakis, Adam Saruwatari, Jojo Callos, Miles Baustista, Jeremy Lookofsky, Bisi Ezerioha, Rodge Sangco and last but not least, Ron and Ed Berganholtz .

CONTENTS

INTRODUCTION

When I was a young teenager, just getting into cars, I was always hungry for knowledge on how to efficiently get more performance from my car. Magazines were my main source of information because they were easy to read and best suited my short attention span. By trial and error (which usually involved blowing something expensive up), I soon found that not everything you read in print is true from a technical standpoint.

After years of blowing things up, at first as a hobbyist, then later as a professional automotive engineer and racer, I decided to write an engine book that would help the hardcore Honda enthusiast. Because there is so much hype and disinformation on the market, I wanted the book to teach some engineering theory as well as just passing out advice. With some basic, easy-to-understand engineering theory, I figured that it would help the readers make educated decisions. This would help them separate the truth from the hype that your average import enthusiast is constantly bombarded with. A little theory can help when purchasing parts as well as tuning them. Theory helps the enthusiast build a solid knowledge base to develop his or her own opinions on equipment and set-ups. This book is meant to be a reference beyond your typical parts selection list.

This is the sort of book that I would have loved to have been able to read when I was younger and just starting out. It was fun writing it. I hope you have just as fun reading it and using the information contained within it pages. Be safe and keep those speed contests where they belong—on the track.

Happy Motoring! —*Mike Kojima*

BASIC ENGINE THEORY 1

So, you are obviously a Honda performance enthusiast or you would not be reading this book. However, because we don't know how much you know, we will cover some basics on how a four-stroke engine works. Some of this will seem painfully obvious to experienced performance nuts, so if you are knowledgeable about the basics of engine operation, please feel free to skip this chapter and read on.

We are covering the basics here because to understand how the latest speed parts work, you first need to understand how an engine works!

This B18C Integra block has been extensively modified to take the stress of racing, you can see the piston bores here.

BASIC ENGINE COMPONENTS

All Hondas, as well as most cars (with the exception of Mazda's Wankel rotary engine), are powered by a four-stroke or four-cycle engine. It is so named because there are basically four strokes to the power cycle: the intake stroke, the compression stroke, the power stroke and the exhaust stroke. It is during this process that air and gasoline are transformed into energy that moves your car.

An engine also has some major parts: the block, the crank, the rods, the pistons, the head, the valves, the cams, the intake and exhaust systems and the ignition system. These parts work in close harmony in an exacting manner to harness the chemical energy in gasoline, converting many small explosions of air and fuel into a rotary motion to spin your wheels and hurl you down the track.

Let's start with the main parts first.

The Block

The block is the main part of the engine that contains the reciprocating components that harness the explosive power of gasoline. The block has bores, cylindrical holes that the pistons slide up and down in. The number of bores equates to the number of cylinders. A four-cylinder will have four bores and four pistons; a six-cylinder will have six bores and six pistons; an eight cylinder will have eight bores and eight pistons and so on. The block also contains passages for cooling water and lubricating oil. Blocks are typically made of cast iron but Hondas have lightweight aluminum blocks. Most Hondas are powered by four-cylinder engines, although some Accords and Acuras have six-cylinder powerplants.

The Pistons

Pistons are cylinders of aluminum that slide up and down in the bores of the block. The top of the bores are closed off by the cylinder head (we will talk about the head later). To make driving power, a flammable charge of compressed gasoline and air contained within the bore is ignited, and the piston is forced down the length of the bore toward the open end of the cylinder, away from the cylinder head with great pressure. This is the basic premise on how an engine works. The piston has rings, which are thin, circular, springy metal seals that fit in grooves around the top

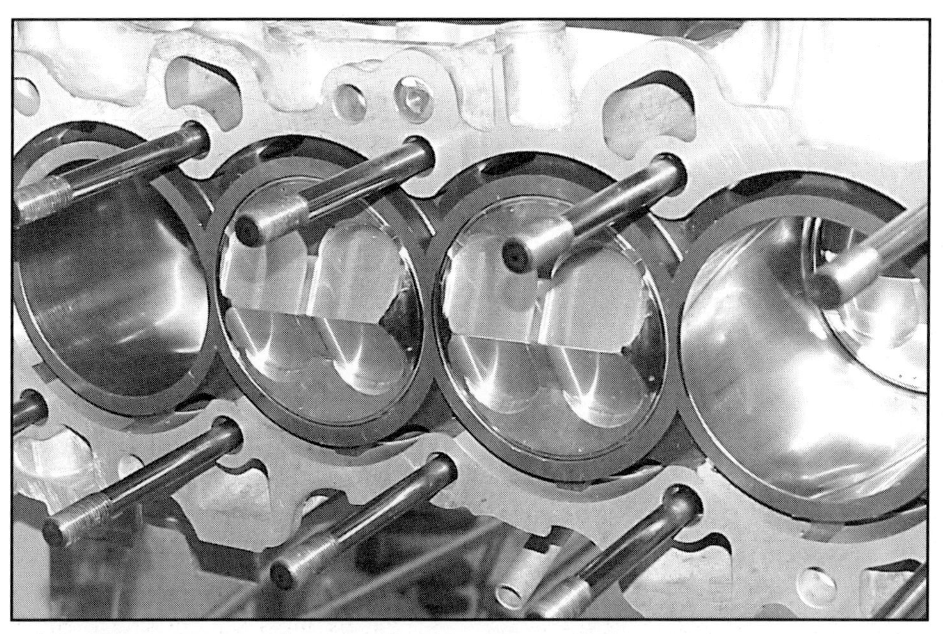

Here is the same block as on page 1 with the pistons installed in the bores. These forged racing pistons have a high dome for more compression than stock.

A close-up view of a very trick forged racing piston. The high dome on this piston is for a high compression all-motor car. It raises the compression by taking up combustion chamber volume.

Here is a view of the block from the bottom showing how the crank and rods are fitted into the block. Those gold-color caps are the bottoms of some very trick titanium connecting rods!

of the piston. The ring's job is to help keep combustion pressure from blowing past the piston. It is this pressure that produces the power. The rings also help scrape lubricating oil off of the cylinder bore walls so it does not burn up in the cylinder. If an engine had no rings, it would not be able to develop enough compression to run, and it would burn up all of its lubricating oil in just a few minutes of running.

Connecting Rods

The connecting rod's job is to transfer the force of the explosion (or "combustion event" for those who are into technical correctness) shoving the piston down the bore to the rotating crank. Connecting rods are attached to the piston on one end and to the crank on the other end. The connecting rod is attached to the piston with a wrist pin, and this end is known as the small end. The other end of the rod is called the big end, as the crank's journals are much bigger than the wrist pin journals. The crank journals are bigger because the crank journal continually rotates at a high speed as opposed to the simple rocking movement at the wrist pin end of the rod. The high-speed rotation requires additional bearing area to prevent the rod and crank from being damaged by friction. The big end of the rod spins smoothly on the journal of the crank on a pressurized oil film or sleeve bearing. On Honda motors, the small end of the rod has a bronze bushing for the wrist pin that is fed by splash lubrication.

Crankshaft

The crank in an engine is exactly like the crank on a bicycle. It transfers a downward driving force, which is the pistons being forced down the bore by the air/fuel explosion, into a rotating motion that can be used to spin the wheels. The crank has offset throws, exactly like your bicycle's crank except the rods and pistons serve the same function as your legs, by pushing the upward throw down as

This Crower 2000cc stroker kit for an Integra B18B motor shows how the rods and pistons bolt to the crankshaft without the block in the way so you can see them better. The relationship of all of these engine parts is easier to understand when it's all out in the open!

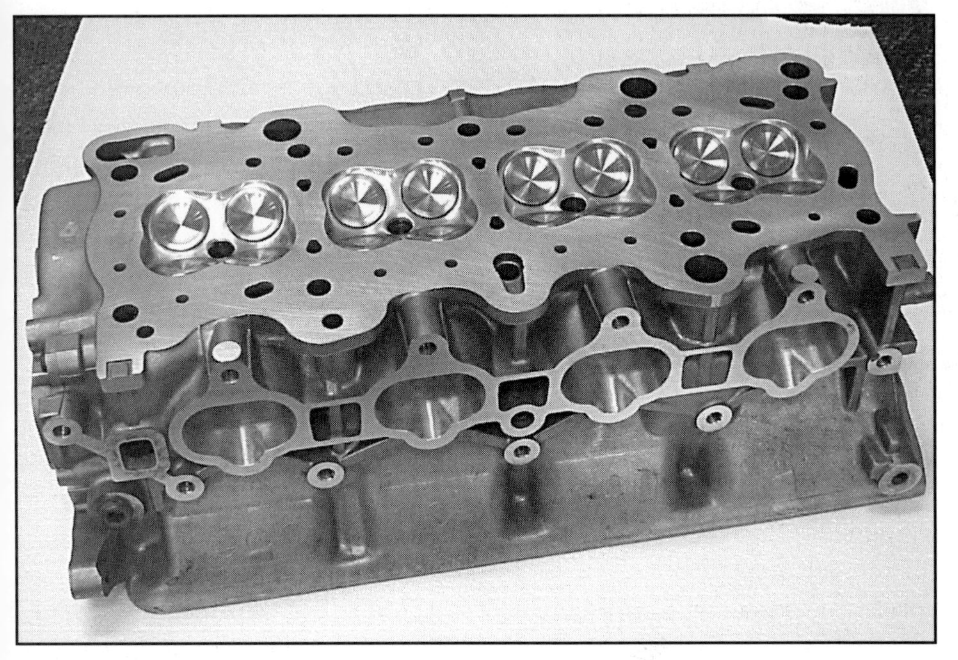

This very trick B18C head has been really worked over by DPR. It features big valves, a modified combustion chamber and porting. We will get into the details of these modifications in a later chapter.

the piston is pushed down the bore by the explosion of fuel and air. This is what makes your car go! After the piston goes down, the crank rotates and the piston is pushed up the bore again until it reaches the top where it can be pushed down again by another explosion of fuel and air. The crank rotates on its main journals on an oil film sleeve bearing just like on the big end of the rods.

The Cylinder Head

Most import engine cylinder heads are aluminum castings that sit on top of the block and contain the spark plugs, valves and the valve train. The head must contain the explosive force of the igniting air/fuel mixture so the explosion of this mixture can only drive the piston down the bore instead of blowing out of the top.

Combustion Chamber—Each cylinder head, directly above the bore, is a depressed concave area called the combustion chamber. This chamber contains the intake and exhaust valves, as well as the spark plug electrode. When the piston reaches the top of its stroke, air and fuel enter the chamber and the spark plug ignites, creating the mini explosion that becomes the power stroke. The cylinder head also has cooling jackets filled with circulated water to help keep the combustion chambers from getting too hot.

Camshafts and Valvetrain

On a modern engine the head also contains the intake and exhaust valves. The intake and exhaust valves are spring-loaded poppet-type valves. The springs hold the valves shut, but allow them to be opened with a push. The intake valves open to admit the explosive mixture of fuel and air into the combustion chamber. They then close to allow the engine to build up compression as the piston, driven by the crank, comes up to TDC or top dead center. When the spark plug ignites the air/fuel mixture and the piston is driven down, and the exhaust valves open near the bottom of the piston's downward travel, allowing the burnt waste gasses to escape to prepare the combustion chambers for the next charge of fresh fuel and air.

This Prelude H22 motor has been modified by JUN. You can clearly see the cylinder head bolted to the block, the intake and exhaust ports, the valves and the piston in this picture. The piston is a forged high dome, high compression model, the valves a big racing stainless steel parts and the intake and exhaust ports have been hogged out and polished. This is a serious motor.

The valves are opened and closed by one or two camshafts, which are basically rods with off-center bumps called lobes that spin in the cylinder head at 1/2 of the crankshaft's speed. The lobes of the camshaft push the valves open and closed so that air and fuel can be admitted with burnt exhaust being expelled. Sometimes the cam can work directly on the valve. Many motorcycles and racing-type engines are like this. Typically the camshaft works the valves through a rocker arm, which is like a miniature teeter–totter. One end of the rocker arm rubs on the rotating camshaft with the other end pushing the valves open and closed.

SOHC—Most sport compact engines, including all Honda engines, are overhead cam engines. This means that the camshaft is contained within the cylinder head on top of the valves. This is opposed to overhead valve engines like low revving, old school domestic V-8's that have the camshaft located in the middle of the block, connecting to the valves with lifters, long pushrods and rocker arms. Overhead cam engines are better for the typical high rpm, small displacement sport compact motor, because they have simpler, lighter, more direct acting valvetrains. These valvetrains work better at high rpm because their lower inertial mass allows them to follow the camshaft's lobes with more accuracy. If the engine has only one camshaft that controls both the intake and exhaust valves it is called a SOHC or single overhead cam motor. The Honda D15 and D16 that is found in the Civic is a classic example of a SOHC motor. The F22 Accord motor is another SOHC Honda power plant.

Many sport compact motors have dual overhead cams, which means that there are separate cams for the intake and the exhaust valves. The advantage with this is that the cam can be placed very close to the valve, allowing the cam's lobes to either work directly on the valves or through a very small rocker arm. This reduces the inertial mass of the valvetrain to a minimum, which helps high rpm operation even more. Just about all of the higher performance Sport Compact motors use dual overhead cam valvetrains also known as the DOHC configuration. The Acura Integra B18B and B18C as well as the Honda B16A (Civic/Del Sol) and H22 (Prelude) are prime examples of DOHC motors.

The Intake System

The intake system consists of the manifold, which is basically a series of pipes that connect the throttle body to the intake ports of the head. The throttle body contains a valve that controls the amount of air the engine can intake, thus influencing its speed and power. When the throttle is shut, the air is very limited, so the engine must idle. When it is wide open, the engine takes in all the air it can so it can produce its maximum level of power.

The manifold usually contains the fuel injectors, which are electro-mechanical valves controlled by the ECU, or engine control unit, a small computer that is the brain of an engine. The ECU controls the amount of fuel being injected into the engine by modulating the open and closed time of the injectors. When the throttle is fully opened, allowing the maximum amount of air possible into the engine, the ECU will command

This JUN intake system is a piece of art suitable for display in a power freaks' museum. It features a big plenum chamber, short runners, big injectors and a big throttle body. All the right stuff for the ultimate low ET racer.

the injectors to stay open longer so they can inject a proportionally greater amount of fuel to create a bigger volume of air/fuel mixture. More air and fuel mixture means a bigger explosion and more power.

The Ignition System

To get the air/fuel mixture burning, the ignition system ignites the flammable mixture by firing a powerful electrical spark across the electrodes of the spark plug. The engine's ECU controls the timing of this spark. The spark is fired as the piston has almost risen to Top Dead Center (TDC) near the peak of the cylinders highest compression. This is the most efficient time to fire the spark. Usually, the timing of the spark advances as engine rpm increases. This is because at higher rpm there is less time for the combustion event to take place, so it must be started sooner in the cycle to maintain proper operation.

The Exhaust System

The exhaust system is simply the tubing that directs burned exhaust gasses away from the motor. The exhaust system consists of the exhaust manifold, the catalytic converter and the exhaust pipe. The manifold collects the exhaust gas from each individual exhaust port in the cylinder head and collects them into a single pipe. This pipe leads into the catalytic converter where poisonous components of the exhaust gas, such as oxides of nitrogen (NOX), various unburned hydrocarbons (HC) and carbon monoxide (CO) are converted to non-toxic CO_2 and water vapor. After the catalytic converter, the gasses flow into the exhaust pipe where they pass through the muffler, which reduces the noise to an acceptable level and out into the atmosphere.

Cooling and Lubricating Systems

Although not directly responsible for motivating your ride but nevertheless deserving an honorable mention as critical parts of a motor is the engine's cooling and lubricating systems.

MSD makes perhaps the best ignition systems available for any price. An MSD is not only highly powered but fires multiple times to make sure the fire gets lit.

DC makes some of the best street headers for Hondas. The quality is pretty much evident here—100% polished stainless steel and TIG welds.

The cooling system circulates a mixture of antifreeze and water throughout the block and head, keeping them from getting too hot from the continuous explosions that they are subjected to internally. The water is pumped out of the block to the radiator, which is a heat exchanger located in the very front of the car, to be cooled down and re-circulated back through the engine.

The lubrication system consists of a pump that feeds oil to all of the bearings located on the crank, and to the valvetrain. The pistons, rings and rod wrist pins usually rely on splash that is created as oil is flung off the rotating crank.

THE FOUR-STROKE CYCLE

Now that you know what all of the basic parts of an engine are and what they do, it's time to understand how they work together in a system. Four-stroke engines use what is called the Otto Cycle or the four-stroke cycle to turn gasoline into power. Manipulation of the four-stroke cycle is essential for obtaining more power from a motor so it is important to know what the different parts of the cycle are and how they affect your power output. Let's jump in and explain the four-stroke cycle.

The Intake Stroke

Let's start with the piston at TDC or Top Dead Center. The intake valve is starting to be opened by the camshaft as the exhaust valve is closing. As the crankshaft turns, the connecting rod starts to pull the piston down away from TDC. The turning crank is also linked to the camshaft by a chain or belt so as the crank turns, the intake valve is opened more and more until it is fully open. The downward traveling piston creates suction in the cylinder so air and injected gasoline from the intake manifold are drawn into the cylinder by this suction. This continues until the piston is all the way to the bottom of the crank's stroke, or Bottom Dead Center (BDC). Because of the shape of the cam, the intake valve is almost totally closed by the time the piston is at BDC. At the end of the intake stroke we are left with a cylinder full of fresh fuel air mixture.

DC also makes a high quality street exhaust system made of polished 304 stainless steel.

ENGINE ID GUIDE

YEAR	MODEL	ENGINE
84–86	CIVIC/CRX/CVCC/Carbureted	EW1, 1.5L SOHC*
85–86	CIVIC/CRX Si/Multi-Port FI	EW3/4, 1.5 L SOHC**
85–87	CIVIC/CRX Si/Multi-Port FI	D15A2/3, 1.5L SOHC**
88–91	CIVIC/CRX Si/ Multi-Port FI	D16A6, 1.6L SOHC***
88–91	CRX HF/Multi-Port FI	D15B6, 1.5L SOHC**
92–95	DEL SOL LX/DX/Multi-Port FI	D16A6, 1.6L SOHC***
92–95	CIVIC DX/Multi-Port FI	D16A6, 1.6L SOHC***
92–95	CIVIC Si/Multi-Port FI	D16Z6, 1.6L SOHC VTEC***
96–2000	CIVIC HX/Multi-Port FI	D16Y5, 1.6 SOHC VTEC-E*
96–2000	CIVIC DX/LX	D16Y7, 1.6L SOHC**
96–2000	CIVIC EX/Multi-Port FI	D16Y8, 1.6 SOHC VTEC ***
2001	CIVIC DX/Multi-Port FI	D17A1, 1.7L SOHC**
2002	CIVIC Si/Multi-Port FI	K20A3, 2.0L DOHC VTEC*
92–95	DEL SOL VTEC/Multi-Port FI	B16A1, 1.6L DOHC VTEC**
86–89	INTEGRA LS/GS/Multi-Port FI	D16A1, 1.6L DOHC***
90–93	INTEGRA LS/GS/Multi-Port FI	B18A1, 1.8L DOHC***
93	INTEGRA GSR/Multi-Port FI	B17A1, 1.7L DOHC VTEC***
94–2001	INTEGRA LS/GS/Multi-Port FI	B18A1, 1.8L DOHC***
94–2001	INTEGRA GSR/Multi-Port FI	B18C1, 1.8L DOHC VTEC***
2002	ACURA RSX/Multi-Port FI	K20A3, 2.0L DOHC VTEC**
2002	ACURA RSX Type S/Multi-Port FI	K20A2, 2.0L, DOHC VTECi**
87–89	ACCORD/Multi-Port FI	A20A3, 2.0L SOHC*
90–93	ACCORD/Multi-Port FI	F22A4/6, 2.2L SOHC**
94–97	ACCORD DX/LX/EX/Multi-Port FI	F22B1/2, 2.2L SOHC**
1995-97	ACCORD V-6/Multi-Port FI	C27A4, 2.7L SOHC V-6*
98–	ACCORD DX/Multi-Port FI	F23A5, 2.2L SOHC*
98–	ACCORD LX/EX/Multi-Port FI	F23A1, 2.2-L SOHC VTEC*
98–	ACCORD LX/EX V6/Multi-Port FI	J30A1, 3.0L SOHC VTEC*
88–91	PRELUDE 2.0 S/Si/Multi-Port FI	B20A5, 2.0L SOHC*
91	PRELUDE Si/Multi-Port FI	B21A1, 2.3L SOHC**
92–96	PRELUDE S/Multi-Port FI	F22A1, 2.2L SOHC**
92–96	PRELUDE Si/Multi-Port FI	H23A1, 2.3L DOHC**
92–96	PRELUDE VTEC/Multi-Port FI	H22A1, 2.2L DOHC VTEC***
97–2001	PRELUDE Si/SH/Multi-Port FI	H22A1, 2.2L DOHC VTEC***
2000	S2000/Multi-Port FI	F20C1, 2.0L DOHC VTECi**
1993-97	ACURA NSX/Multi-Port FI	C30A, 3.0L DOHC VTEC**
1998-	ACURA NSX/Multi-Port FI	C32A, 3.2L DOHC VTEC **

TUNING SCALE

* Parts are limited, low power output

** Parts are available, mid-power output, mild applications

*** Parts are abundant, high power output, mild to wild applications available

The Compression Stroke

Next, the piston begins moving up, pushed by the crankshaft and the connecting rod. Now the intake valve is fully closed and as the piston is forced upward, the air/fuel mix is compressed. This compression forces the fuel and air molecules closer and closer together until they become a highly reactive explosive mixture; the closer the proximity of the molecules, the easier it is to initiate an explosion. When the piston nears TDC again, the ignition system fires the spark plug, which triggers an explosion in the cylinder.

The Power Stroke

By the time the piston is at TDC the explosion of fuel and air in the tightly contained cylinder is well under way. The heat and pressure of the explosion rise rapidly and the piston is pushed strongly back down the cylinder with great force. This is the driving power that spins your wheels and propels you down the track. As the piston is pushed down the bore, the cylinder pressure starts to decrease as the volume of the cylinder increases. As the piston nears the bottom of the bore, the camshaft starts to open the exhaust valve.

The Exhaust Stroke

Nothing too exciting goes on here. As the piston once again starts to go up from BDC, the exhaust valve opens and the burnt exhaust gas is forced out of the cylinder by the rising piston into the exhaust manifold, through the catalytic converter, down the exhaust pipe, through the muffler and out. By the time the piston is back at TDC the exhaust valve is almost closed and the intake valve is starting to open. The cycle is about to start again.

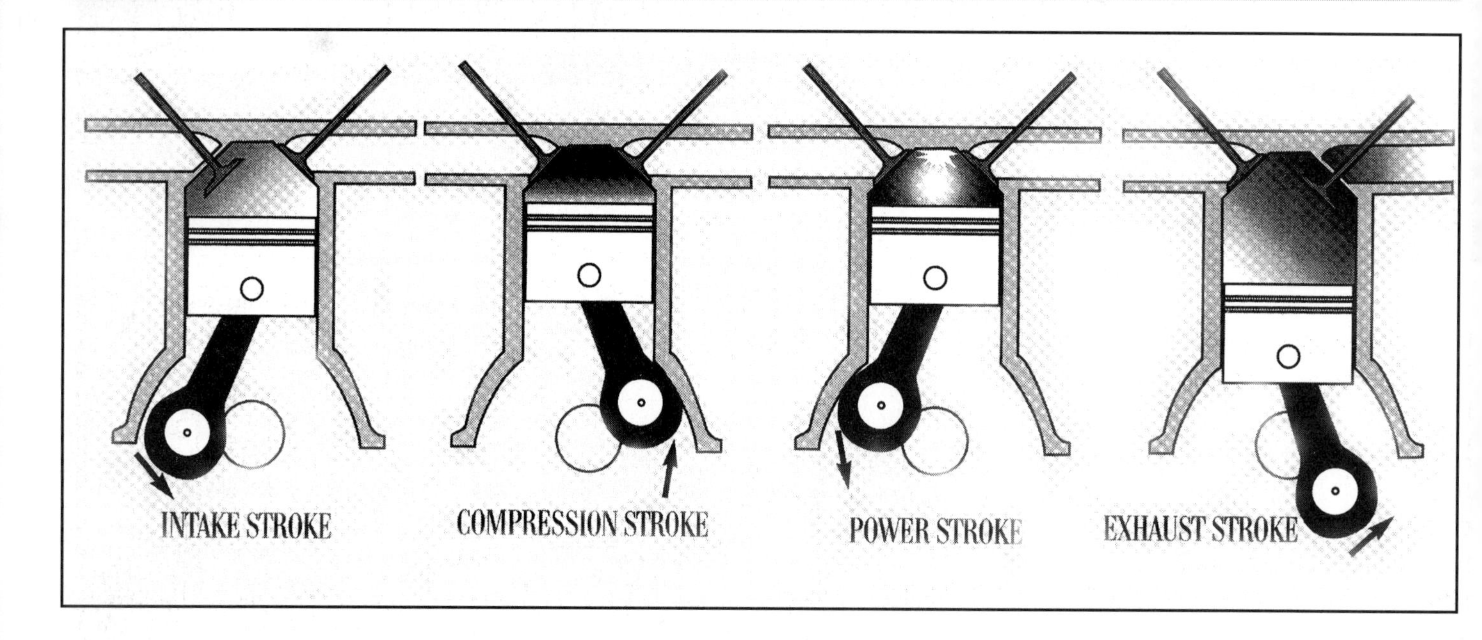

A four-stroke engine has one power stroke in each cylinder for every two revolutions of the crank. Since a camshaft has one bump on each lobe, it is driven at 1/2 the engine speed, so the cam will open the valve at every other revolution of the crank as necessary.

Imagine this stuff happening at 8000 rpm! The cycle would be repeating itself at about a rate of 60 times a second in each cylinder. At that speed it is easy to imagine a more uninterrupted flow of power coming from this frenetic system.

Modifications

Now when you are going to attempt to modify your motor for more power, you must remember that you can never get something for nothing. The factory usually does a remarkable job of designing an engine that can meet the many demands made of a modern powerplant. Remember, most of the guys that work for the manufacturer, especially Honda, are not dummies, they just have to play in a game with many rules!

A stock engine's design goals are not solely to make the maximum amount of power possible. Stock engines must run smoothly and quietly, last a long time, get maximum fuel economy, run on 87 to 92 octane without blowing up, make as little pollution as possible and most of all, they must be cheap to mass produce.

As an enthusiast driver, you must be willing to sacrifice some of these design guidelines to get more power, which most of you are willing to do. To go faster, you must always burn more fuel. In the following chapters, we will talk about the modifications you can do to your Honda from headers to turbos, and how they work so you can build your Honda motor from mild to wild with some acquired expertise and a minimum of mistakes.

BOLT-ON MODS

2

Everyone has to start somewhere in the path to building a fast Honda. Most of us don't have the money or the knowledge to build a copy of Stef Papadakis's tube frame 8-second monster Civic right off of the bat. For most of us, some of the first stops on the road to high performance are parts that can be simply bolted on. Bolt-ons are usually installed with a minimal of hassle, usually without having to open up the internals of the engine. Bolt-ons are simple mods that can usually be done by a beginner with a basic set of hand tools.

The good news is that these simple bolt-ons will produce a noticeable gain in horsepower, enough to get a few car lengths more over a stocker in a quarter-mile heads-up race.

Once you have decided that you want more power and understand the compromises that you must make to get it (more noise, rougher idle, worse gas mileage, depending on how far you go) you can decide what parts to get.

This Civic Si features many of the hot bolt ons: a DC Sports DAC air intake system, a DC Sports Header and an adjustable fuel pressure regulator for tuning.

THE EXHAUST SYSTEM

A replacement cat back (which means everything from the rear of the catalytic converter back) exhaust system is perhaps the best place for a beginner to start. Unlike some hard-core modifications, the exhaust system usually will not affect mileage, emissions or driveabilty very much. Because of this, the exhaust system is one of the first areas we recommend to replace. The cool looks and sound of a performance exhaust are also appealing.

Since an engine is a glorified air pump, much of its efficiency is based on how easily exhaust gas can get out of the cylinders. Restrictions to this outward flow are called *pumping losses* by engineers. Obviously, the least amount of obstruction will reduce the pumping losses to a minimum, freeing up more power to the wheels.

The exhaust system is, obviously to some, the piece of piping that directs the car's exhaust stream from the exhaust manifold (the branched collector that gets the exhaust from each individual exhaust port and brings it into one pipe) to the tailpipe. To get there, it must first pass through the catalytic converter (the emissions device that converts poisonous exhaust emissions to H_20 vapor and CO_2), through the muffler and out the back of the car.

The purpose of the exhaust system is to contain the noisy, hot, toxic exhaust stream. With no exhaust system, a car would be painfully loud and spew lots of potentially deadly fumes, poisoning our atmosphere. To prevent this, the exhaust system must quiet the engine's noise and remove the byproducts of combustion before discharging the exhaust gasses into the air.

The exhaust system that comes stock on your Honda or Acura was not designed for power and cool looks as a priority. Honda's engineers designed the exhaust to be as quiet as possible, be 100% emissions

This DC Sports all-stainless fully polished exhaust system has superb construction quality as well as good performance.

JUN's titanium muffler is perhaps the most esoteric of all of the mufflers on the market. Besides having a good free-flowing design, its titanium construction ensures that it will not rust and be extremely light. As for cost, if you need to ask, you can't afford it!

compliant, to last for as long as the vehicle's warranty period, and to be as cheap as possible to produce. These are the attributes that 95% of the motoring public deems important, but not the performance enthusiast. To you, the roar of a tuned engine is music to your ears. To some geeky accountant, it is terrible noise. Since your car is designed to appeal to what most customers want, and you are in the minority of car buyers, your exhaust ends up getting optimized on the quiet side of the decibel scale, compromised on the performance end.

The Muffler

The key part of your exhaust system is the muffler. The muffler is the can at the end of the exhaust whose job it is to make the exhaust quiet. To do this, a typical stock muffler must have an intricate and labyrinth-like internal flow path to help slow and cool the hot, vibrating exhaust gas. It contains baffles that cause the exhaust flow to reverse direction and mix. These baffles are great for reducing noise but are not good for power. The network of baffles that redirect and intermix this flow are very restrictive, impeding the flow of gasses from the car, which directly affects power.

To produce the most amount of power, an exhaust system should restrict exhaust flow as little as possible. Restriction hampers the burned exhaust gasses from exiting the engine, causing some charge dilution with the incoming fresh air/fuel mixture. This causes a loss of power. Greater restriction also results in backpressure, which makes the engine work harder to pump the exhaust out of the cylinders. This is what I meant by pumping loss discussed previously.

Some stock mufflers produce up to 18 psi of power-robbing backpressure. A well-designed performance exhaust typically has about 2–6 psi of backpressure. For comparison sake, an un-muffled straight pipe as found on an off-road racecar usually has 1–3 psi of backpressure. Obviously, you would never want to run an un-muffled pipe on the street unless you want to get to know your local law enforcement real well.

To get the least amount of backpressure, most of the good high performance mufflers available today have what is called a straight-through design. These types of mufflers quiet the exhaust by the absorption of high frequency vibrations with a fluffy packing material, usually consisting of stainless steel mesh and heat-

A typical high-quality absorption muffler has blow-out resistant stainless steel wool surrounding the center pipe, then fiberglass packing. This two-stage type of packing lasts much longer than just simply having fiberglass, which can blow out fairly quickly.

This Mugen muffler features stainless wool surrounded by fiberglass packing.

Here is what a louvered core looks like from the inside. Even though this design is straight through, it still can have a lot of backpressure. This type of muffler is best avoided.

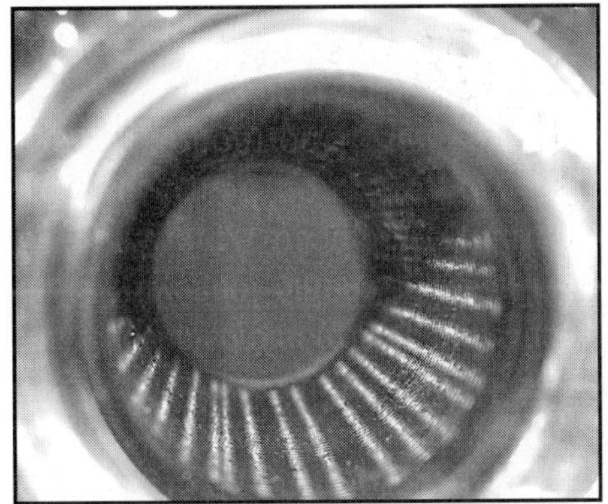

Here is a perforated core-type muffler. This type of core is the lowest possible restriction of any muffler and has the same restriction as an open pipe! In my opinion, these are the best performance mufflers you can buy.

The ultra-trick JUN titanium muffler has a high-flowing perforated core.

resistant ceramic fibers. They typically have an inner core that is straight through with no baffling at all, much like a straight pipe with many small holes. The pipe is louvered or perforated when it passes inside the muffler's shell, allowing the radiating sound energy to pass through the holes to the packing where it is absorbed and damped. While this is going on, the exhaust gas flows out of the car unimpeded. You can see straight through these types of mufflers. The louvered or perforated core is usually wrapped with either fiberglass padding, hence the old school term "glass-pack" or on the better mufflers, stainless steel mesh and ceramic wool to help further absorb the sound. On straight-through mufflers the longer the muffler, the quieter it is. The length usually has no effect on the backpressure, just the noise output. These mufflers work in the same manner as silencers do on guns. If a silencer impeded the bullet's travel, you would definitely have problems with your gun!

Recommendations—It is best to avoid those types of mufflers that have a louvered core. Many old school glass packs suffer from these. The louvers generate quite a bit of backpressure because they stick into the exhaust stream and create turbulence. Even though these mufflers are a straight-through design, they can have more backpressure than a baffled stock muffler! When buying a straight-through muffler, look for one with a perforated core if you are interested in producing more power. A good, properly sized, perforated-core, straight-through muffler will add

Here is a high-quality Magnaflow pre-silencer and muffler. The Magnaflow has a perforated core and all stainless steel construction.

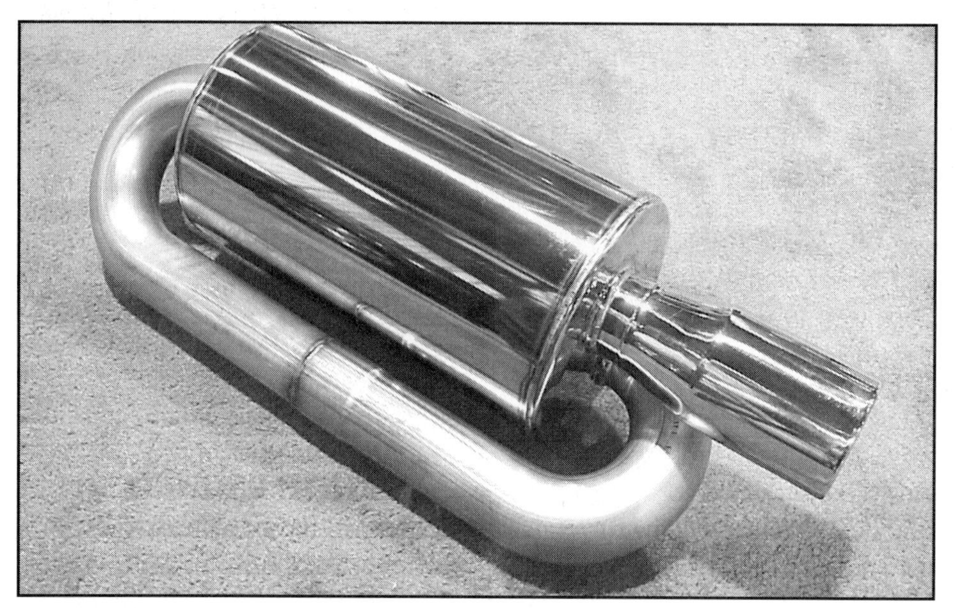

Hi Tech Exhaust has an unusually designed muffler. It is a perforated core but the exhaust makes two passes through the muffler body, eliminating the need for a pre-silencer.

only about 1–2 lbs of backpressure to your exhaust system. Mufflers like the Walker Ultra Flow, Thermal, Apexi, Borla, Tanabe, Edelbrock or Magna Flow are examples of a good low backpressure mufflers with an absorption design. Many pre-made exhausts like Apexi, Tanabe, Greddy or HKS also have a free-flowing absorption design.

Turbo Mufflers—A very popular, though somewhat dated performance muffler that has seen better days is the Turbo-style muffler. This is a less-restrictive version of a stock-like reverse-flow muffler. A few years ago, these were good-flowing mufflers and very popular with the V-8 crowd, but the current generation of perforated core, straight-through absorption-type mufflers flow much better.

Exhaust Tips—Many enthusiasts like to add a big exhaust tip. These look great but do nothing to increase power. Some people just add the tip without adding a performance exhaust, which is sending a signal that you are just an amateur and not a real player. If you are on a tight budget, don't get a big exhaust tip until you get the performance exhaust system to go with it. Some big tips feature resonated cores, which quiet the exhaust note by a few decibels. These are usually identified by perforated or meshed inner pipes. These big tips are actually functional, so you won't be teased or put down if you show up at the track with one of these.

A disadvantage of a straight-through muffler is that it is louder than a reverse-flow type. Usually, a straight-through muffler needs a small sub muffler or a resonator to keep the exhaust quiet. A resonator is usually a small, perforated-core glass pack placed somewhere in between the catalytic converter and the main muffler. Like the rear located main muffler, the longer the resonator, the better for noise reduction. A Walker Magnum Glass-Pack is a good muffler to use as a resonator. Almost all of the pre-made performance exhausts feature resonators.

When designing your own custom exhaust, it is important to remember to make it as quiet as possible. You might think the racy sound is cool, but your local police will not.

The Exhaust Pipe

To save costs, your stock Honda exhaust is made with small-diameter,

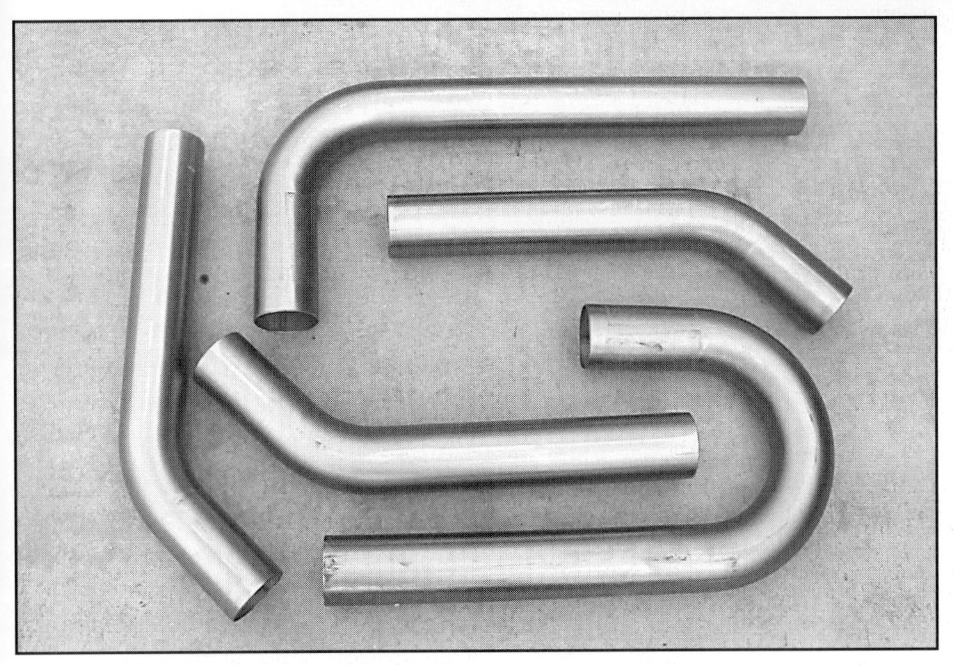

Here is an assortment of 304 stainless mandrel bends offered by Magnaflow. A decent fabricator can make a nice exhaust from these sections by cutting the radius of the bends needed and welding them together

crush-bent pipe. Crush bends are easy and inexpensive to make, which makes them ideal for mass production. However, crush bends can reduce the flow of a pipe by up to 50%. The typical exhaust system made by a standard muffler shop is also crush bent. The best exhaust systems, like most pre-made exhaust systems, come with mandrel bends. Mandrel bending is done with a special machine that uses a non-crushable insert, called a mandrel, that goes into the pipe while bending to prevent it from becoming crushed. The bend is smooth, which helps with exhaust flow. If you are making your own exhaust you can buy pre-made mandrel bends from Burns Stainless, Magnaflow, Kinsler or Bassini.

Pipe Diameter—Some self-proclaimed motor gurus state that you should not run an exhaust pipe that is too large in diameter because engines need a certain amount of backpressure to run correctly. Although the statement about size is correct, the assumption about engines needing backpressure is not. An engine needs to have the lowest backpressure possible to produce the maximum power by keeping pumping losses low. An exhaust pipe that is too big causes power loss, especially low-end torque, because a big pipe has less exhaust stream velocity than a smaller pipe. Velocity is essential to get the best scavenging effect from tuned headers, which we will discuss in more detail later. To simplify things, if the exhaust gas flow is kept high with good velocity, a vacuum can develop behind the closed exhaust valve, allowing even better scavenging when the exhaust valve opens on the next exhaust cycle. Good scavenging is even more critical on valve overlap, the part of the 4-stroke cycle where both the intake and exhaust valves are open.

If the exhaust pipe diameter is too large, the flow will be sluggish with low velocity and the scavenging will not be so good. Remember that a good exhaust has low backpressure and high velocity. The only possible exceptions to this rule are for turbocharged or nitrous motors. It is almost impossible to put an exhaust that is too big on a turbocharged engine, because the turbo depends a lot on the pressure differential across its turbine to make power. A turbo engine can have an exhaust gas volume about 1.5 to 2 times more than an equivalent displacement, naturally aspirated motor. NOS motors also have a pretty high exhaust volume and require a bigger exhaust if they are to be optimized for NOS operation.

Some basic exhaust pipe diameter guidelines for Honda and Acura motors are as follows:

D15, D16, B16A, ZC: 2 inch
B18B, B18C: 2.25 inch
H22, H23: 2.5 inch

Add 1/2-inch to the pipe diameter to optimize for NOS use. For turbo motors, 2.5 inches is the minimum size diameter pipe that you would want to run, even for the smaller motors. For the B18 1800cc and bigger H series motors 3 inches works well.

If you don't feel like making your own system, a reputable company like DC Sports, HKS or Greddy makes well-engineered, pre-made exhaust systems. These systems have a good balance of low backpressure, properly sized tubing and low noise.

Net Power Gains

The stock exhaust systems on most of the newer Hondas are so well designed nowadays that just switching to a high performance cat-back exhaust often does not allow for huge

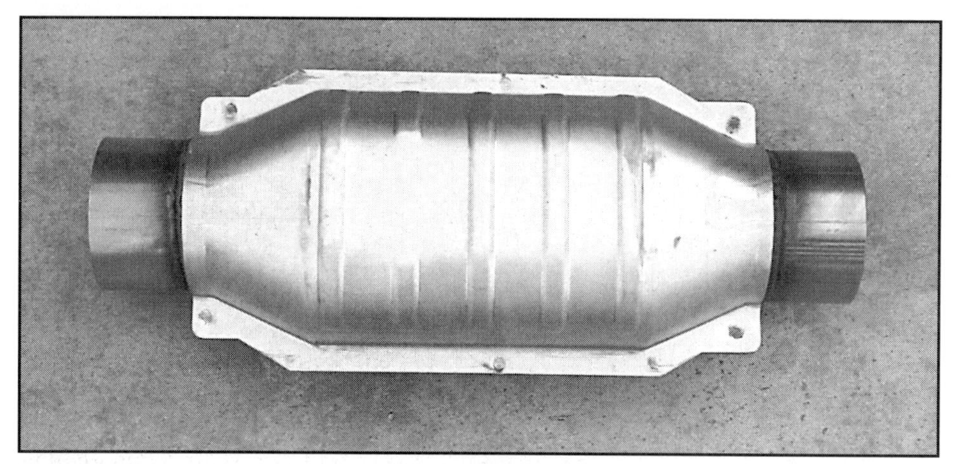

This Magnaflow, high-flow cat has hardly any restriction and can be used even on built motors with little or no power loss. Even the stock Honda cat is excellent. The cat should only be removed on actual race cars, not cars driven on the street.

gains of power. With the exception of turbo cars, you can usually expect only about a 2 to 5 hp increase at the wheels on a stock car. The power gain will usually be the greatest from just below the torque peak to the redline. On cars that are modified with headers, intakes, cams, headwork, etc., you can expect bigger gains with the addition of a cat-back exhaust due to the higher flow volume that these mods produce.

As turbo cars are very sensitive to backpressure, you can expect much larger gains with them, especially if the boost is turned up. A free-flowing exhaust usually allows the turbo to spool faster also. A turbo car usually gains from 8–30 hp at the wheels, depending on how radical the turbo system is and how high the boost is turned up over stock.

Hollow Cats—Whatever you do, do not remove or hollow out the catalytic converter on your street car. The monolithic, straight-through design of Honda-Acura catalytic converters is usually quite free flowing, producing only a pound or two of extra backpressure. A hollowed cat can actually hurt power as the empty box can cause flow stagnation, which effectively shortens the length of the moving gas column in the exhaust pipe. The empty box can also reduce important flow velocity. This is usually felt as a loss in bottom end power, although power throughout the range is affected. Oh, and it is also illegal.

As the number of vehicles on our highways grows every year, we must all do our part to manage pollution. If every last bit of power must be extracted for off-the-street sanctioned racing (meaning only on the race track), then the cat can be removed and replaced with a length of pipe, rather than gutted. When you're done racing, reinstall the cat and drive home.

For most naturally aspirated motor build-ups, the stock catalytic converter has more than adequate flow. Very little can be gained by going to an aftermarket cat. If you have a supercharged or a turbocharged car, Random Technology and Magnaflow makes replacement cats with 3-inch or even larger inlets and outlets. When choosing a high-flow cat, it is important to get one with a brick (the slang for the catalytic matrix) with a large frontal area as well as a big diameter inlet and outlet flange. A larger cross-section brick area is as important as inlet and outlet size when it comes to reducing backpressure.

Headers

Headers can produce substantial amounts of power on a motor with very few negative compromises. They are a rare, win-win modification with hardly any negative trade-offs. This makes them a mod that is essential for

This DC Sports header is a beautiful work of art, crafted from polished 304 stainless steel. This header is a Tri-Y design for a broader powerband.

This Apexi header is a Tri-Y type featuring merged collectors, see the carefully mitered tube junctures? This is a good smooth merge.

any serious engine build-up.

Due to space, cost and emissions-driven, catalytic converter light-off reasons, most Hondas come with a crude, cast iron, log-type manifold stock from the factory. A log manifold is simply a tube with stubby legs connecting the exhaust ports to the main tube. This is good for conserving heat to quickly light off a catalytic converter during cold starts, and it is compact, preserving valuable underhood space in today's crowded engine compartment. However, a log manifold is detrimental for power production. Honda manifolds are much better than the ones found on your typical domestic, coming with cast iron crude approximations of headers. This is better than a log but more power can be made with a real tubular header

A header is an exhaust manifold fabricated from tubular sections of pipe. Full-radius mandrel bends are preferred so the pipe's tight radiuses will not be crushed down. Each individual exhaust port is treated to its own separate primary runner instead of merely dumping into the shared main pipe of a log manifold. The near-equal length primary pipes converge at a single, larger diameter point called a collector. The collector then leads to the main exhaust pipe.

An old hot rodder's tale is that headers produce more power by reducing backpressure and because the long individual runners prevent the exhaust of one cylinder from blowing into the next one, contaminating the charge on overlap. While this is partially true, it is not the primary reason why headers produce more power than the stock manifold on modern cars.

How They Work—Headers make more power by primarily using *resonance tuning* to create a low-pressure, reflected wave, rarefaction pulse during the overlap period. The overlap period is in between the end of the exhaust stroke and the beginning of the intake stroke (remember our 4-stroke cycle from Chapter 1?), where both the intake and exhaust valves are open at the same time for a few degrees of crankshaft rotation. Engine designers use overlap to help the engine breathe better.

To work properly, a header must first exploit the inertial force of the out-rushing exhaust gas. This rapidly moving, high-mass, high-pressure pulse helps create a low-pressure suction in its wake to pull burnt exhaust gas out of the cylinder. This first negative pressure wave helps evacuate the cylinder of burnt exhaust as the piston nears TDC and slows down.

To get the best breathing and to help pull as much fresh air/fuel mixture into the cylinder during the overlap period as possible, it is best if a low-level vacuum or rarefaction can be created and maintained past the initial low pressure wave in the primary pipe. A well-designed header can use acoustic energy to maintain low pressure near the exhaust valve during the overlap period.

Tuning Headers—The way a header is tuned is much like how an organ pipe is tuned. The optimal length used is the one needed for the primary pipe to have a fundamental note corresponding to the time when the exhaust valve opens. When the exhaust valve opens, a high-pressure pulse of hot expanding exhaust gas travels down the exhaust port at approximately 300 feet per second. This wave of hot, moving, high-pressure gas has mass and inertia of its own which pulls a suction or a

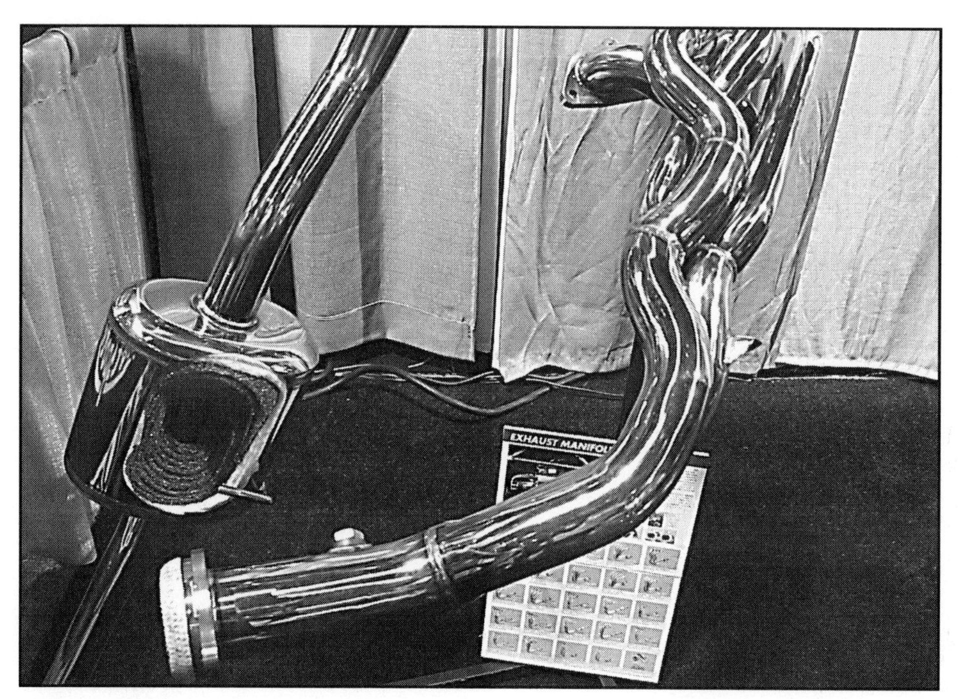

Mugan's S2000 headers feature Tri-Y construction, equal length primaries, merged collectors and polished stainless steel construction—nice pieces! The Mugen headers are one of the only ones to get a significant power gain from the well-engineered 240 hp F20C1 S2000 motor.

low-pressure rarefaction behind the pulse.

Depending on the engine and what is done to it, the pulse can have a positive pressure of anywhere from 5 to 15 psi with the low-pressure rarefaction behind the pulse being anywhere from 1 to 5 psi of negative pressure. As this low-pressure rarefaction is several milliseconds behind the initial high-pressure pulse, it can be exploited to help suck residual exhaust gases out of the cylinder toward the end of the exhaust stroke as the piston approaches TDC. The buildup of this negative pressure and its timing in the exhaust stroke is closely associated with the primary pipe's length and diameter, just like an organ or other musical instrument.

Overlap—This sort of evacuation affects the latter part of the exhaust stroke, making it more efficient by helping suck the last bits of residual exhaust gas out of the cylinder, improving its condition for the next intake stroke. It is also important to continue to maintain a low-pressure condition near the exhaust valve as the piston approaches and passes TDC on the exhaust stroke while the exhaust valve starts to close and the intake valve starts to open. This point where both the intake and exhaust valves are open at the same time is called the overlap period. This is where the last bit of exhaust is expelled from the cylinder while fresh air/fuel mixture is introduced at the same time.

During the overlap period, the piston is starting to slow down as it approaches TDC and gets ready to reverse directions. To maintain good scavenging, a negative pressure must be maintained near the exhaust valve to help continue to suck stale exhaust gas out of the cylinder to make room for fresh fuel and air. As the main column of high-pressure gas is almost out of the end of the header's primary tube, the pressure near the exhaust valve starts to rise again. All is not lost however.

As the pulse of high-pressure, high-energy gas leaves the end of the primary tube and is diffused in the larger diameter header collector, a reflected pulse of sound energy, just like a musical note, is generated, much like that of an organ. This is exhaust noise. This reflected sonic pulse travels down the exhaust pipe at sonic speed, which is usually around 1100–1900 feet per second in thin hot exhaust gas, causing a slight rise in pressure at the valve. The wave is then reflected back down toward the open end of the primary pipe, trailing a rarefaction behind it. If the pipe is of proper length and diameter, this reflected wave could be exploited to lengthen the amount of time that the condition of low pressure exists around the exhaust valve.

These phenomena are harnessed by header designers to tune the pipe to help get the maximum amount of burnt gas out and to help pull the most fresh fuel in. Of course, because a header is tuned like a musical instrument, a header can only be optimized to produce the greatest scavenge-improving vacuum in a band of several hundred rpm.

Collectors—The way in which the primary pipes gather together is also important. This area of convergence, or *collector*, is critical for proper header function. It must be of larger diameter than the primary tubes because it must be large enough to acoustically represent the end of the pipe for tuning reasons, and it must be big enough to support the flow from all the cylinders without creating excessive backpressure. Usually, the

A DC Sports 4-1 header is more of a race set up designed for peak top end power.

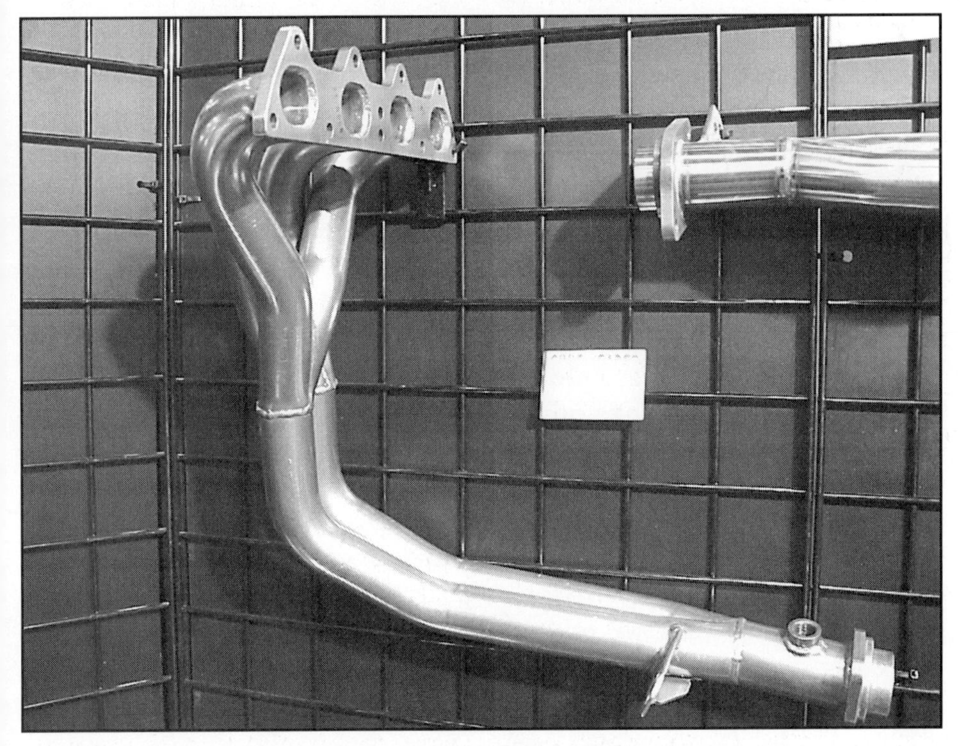

Comptech also makes a Tri-Y header for the venerable Civic D-series motor.

collector is just a junction where all of the pipes are stuffed and welded into a larger pipe that may or may not neck down into the final size of the exhaust pipe. A well-designed collector pairs cylinders opposite in the firing order with each other so an exiting pulse from one cylinder will not hamper the evacuation of the next cylinder. Adjacent cylinders in the firing order are kept separate so that the exiting pulse of one cylinder cannot contaminate the next cylinder that may be on the overlap part of the power stroke. In a typical inline-4 cylinder, that would mean pairing cylinders 1–4, and 2–3.

The best collectors are called *merged collectors*. This is a collector where the two opposite cylinders are paired together in a smooth taper before being introduced to the flows of the other cylinders. Merged collectors usually produce a wider powerband and sometimes more top end power. Not too many production headers are merged due to the difficulty in fabrication, but many merged collector headers are found on real racecars.

Tri-Y Headers—Many headers presently available for Hondas are of the Tri-Y design. For street cars, Tri-Y's are usually the best, as they are forgiving to camshaft design and other tuning factors that the header builder has no control over, unlike a real race-car designer who knows exactly what is in his engine. Tri-Y's also promote a wide powerband. A Tri-Y design pairs the opposite cylinder in the firing order together in a short "Y" and then brings the two pairs of "Y's" into a single collector, hence the name Tri-Y.

When a pulse travels down the primary of a Tri-Y header to the collector, it mostly goes down the main branch of the primary. When it reaches the collector, the reflected wave also travels back up the main primary to the exhaust valve and back out again. However in a Tri-Y, the branch that goes up to the opposite cylinder acts like an interference branch, since the exhaust valve is closed for that cylinder. This creates a pulse and an assisting wave of its own, slightly out of phase with the

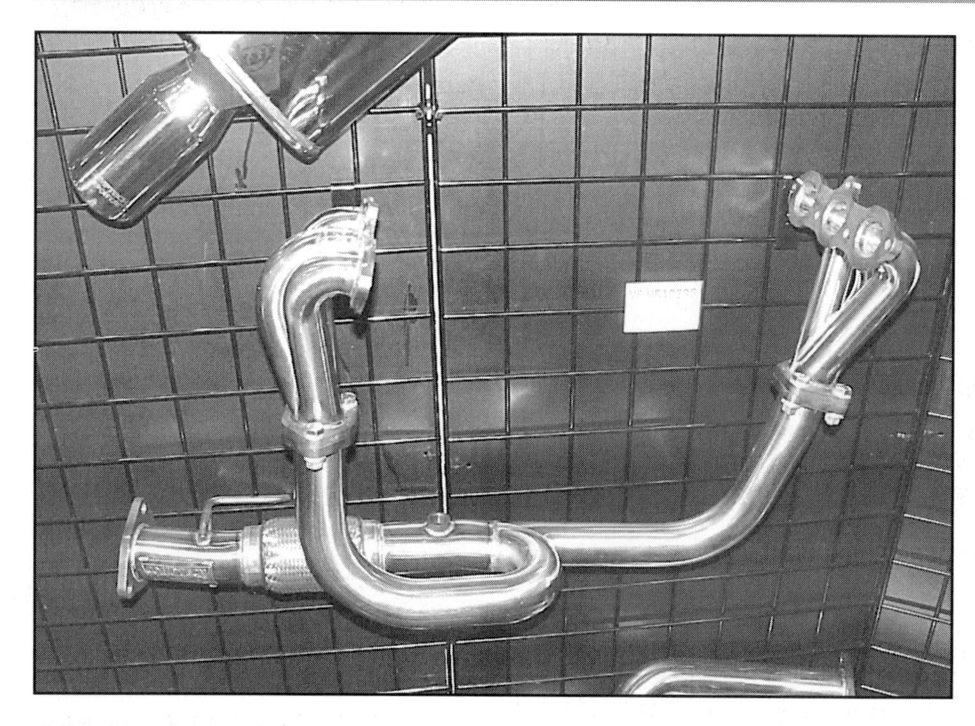

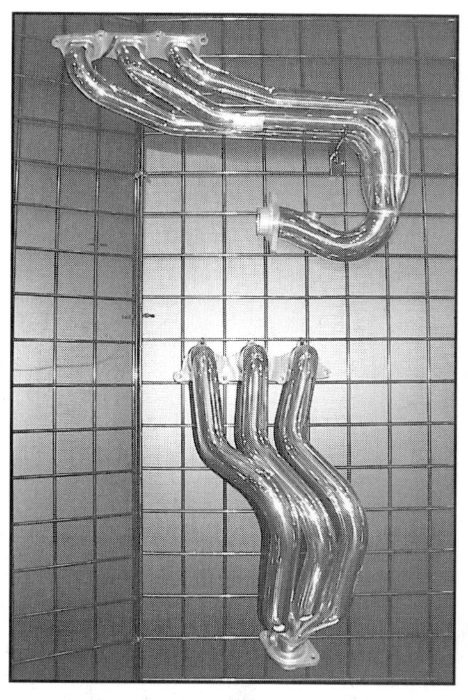

Comptech makes headers for the F-series V-6 J30A Accord and the C-series NSX C30A and C32A motors. Not too many companies make parts for these cars.

main pulse and wave. This widens the bandwidth of rpm that the additional scavenging is effective by making the pipe less sensitive to rpm induced pitch.

The pipe becomes "in-tune" for a longer band of rpm, widening the engine's powerband at the expense of slightly reducing peak power over a 4-into-1 design. Since some of the pulse's energy is dissipated in the interference branch, the main pulse is not as strong and the scavenging effect is not as total for the Tri-Y. Peak scavenging efficiency is compromised for having good scavenge over a wider range of rpm. That is why many full-race engines, where peak power is important, use 4-into-1 designs, while many headers that are designed for best driveabilty, like street engines or rally engines, use Tri-Y headers.

Basically, the majority of street performance cars are better off with either a Tri-Y or a 4-into-1 with either long runners, small runner diameters or both, preferably with a merged collector.

Smog-Legal Issues—For street cars it is essential that the headers you purchase have provisions for all of the vehicle's stock O_2 sensor and any emission controls that the vehicle originally had fitted to the exhaust manifold, although Hondas have not had any emissions fittings on the exhaust manifolds since 1984. Most modern emission controls do not rob any wide-open throttle horsepower anyway. The common EGR valve, which reduces toxic oxides of nitrogen, closes and has no effect at wide open throttle. Removing these controls does not help power and pollutes the air. This is not good for a street car as we must all do our part to help keep our planet clean.

On a more sinister note, on the new OBD II cars (1995 and later), if any of the emission controls are not hooked up, the car's ECU will start recording the error codes generated. In many states, you cannot register your car if the ECU has uncorrected error codes in its memory. There are many in our government that would like to closely monitor these codes and restrict our driving privileges if these codes can be found in our ECUs, all the more reason not to mess with emissions.

Just because your headers have provisions for the O_2 sensor, don't assume that it is street legal. Due to the intelligence of some of our local government agencies, unless an aftermarket part is CARB-approved with a CARB EO number, it is not legal in some states no matter how clean the gasses coming from the tail pipe are! So if no smog certification hassles are important to you, either check your local laws before installing or make sure that the part you buy has a CARB EO number. Your parts dealer should be able to answer that question.

You can expect a power gain of about 3–10 hp at the wheels from a well-designed header on most Hondas, depending largely on how

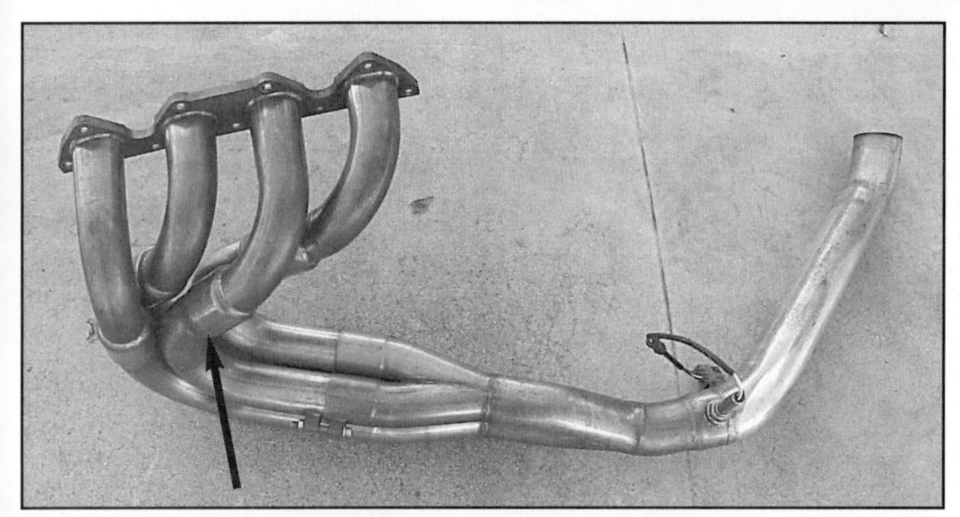

The High Tech Exhaust semi-custom, anti-reversion header is one of the carefully guarded secrets of the quickest All Motor class cars. The gold color of the header is the trick, heat-resistant 321 stainless steel construction. Although expensive, the power gains from this header are phenomenal and the construction is exquisite. The arrow indicates the anti-reversion chamber.

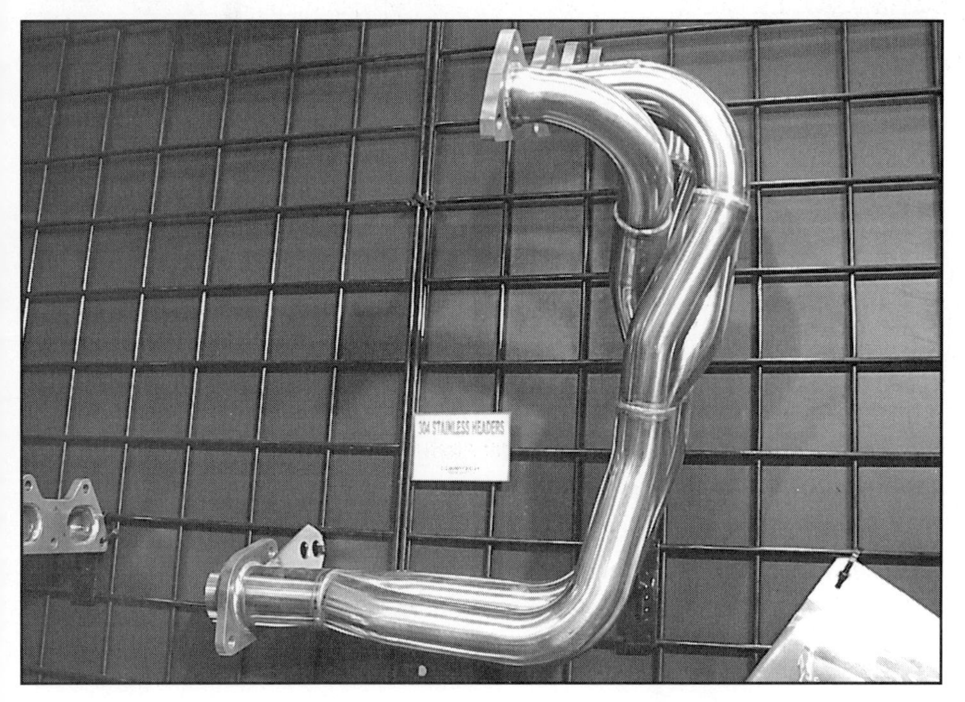

The Comptech B18C header has stepped primary tubes for an anti-reversion effect.

modified the engine is. Engines in a higher state of tune, especially ones with cam changes, usually see a greater power gain with headers. The Honda factory exhaust manifolds are a pretty good design, so you don't get as large gains with headers as other makes of cars can. If you maintain the same driving habits and don't exploit your newfound power too frequently, you can expect better mileage with a header due to the improved pumping efficiency and more complete scavenging it produces.

Selecting Headers—When selecting headers, keep in mind that shorter primary runners and bigger diameter primary runners are better for top-end power. This has to do with the tuning of the pipes' fundamental note for reflected wave tuning and the travel time of the main initial exhaust gas pulse. Just like a piccolo is a higher pitched instrument than a clarinet, a shorter, fatter primary pipe is better for higher rpm. Conversely, a longer and/or thinner in diameter primary tube is better for lower rpm for the same reasons as above. Camshaft design and the duration of the exhaust cam are a large factor in header design. Generally, the later the closing point of the exhaust valve, the shorter the header primary pipes must be.

When buying a header, look for thick-wall mild steel tubing, at least 16 gauge but preferably 14 gauge. Rust and heat-resistant, ceramic-coated or stainless steel primary pipes are preferred for longer life. Look for thick flanges also as these will resist exhaust leaks and last much longer.

If you are lucky enough to own a VTEC B16A or B18C, Honda has an excellent factory header for you. This is the manifold from the Japanese Integra Type R or the JDM manifold, as some people call it. It is a high-quality, 4-into-1 tubular stainless steel factory header! H22 VTEC Si and SH Prelude owners can also buy the JDM Type S Prelude manifold. This is a Tubular Tri-Y factory stainless header. Many import companies sell brand-new JDM Honda parts.

If you prefer to buy an aftermarket header, Mugan, DC Sports, Hotshot, Stillen, RSR, Brospeed, Apexi, Greddy, Comptech, Airmass and many others make good quality headers for most popular Honda models.

Anti-Reversion Headers

These innovative headers deserve a category of their own. Anti-reversion

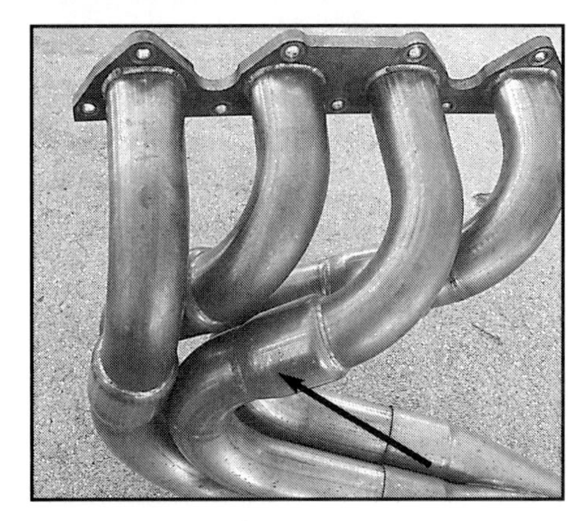

Here is the anti-reversion chamber in the primary tube of a High Tech Exhaust race header. This chamber modifies the timing of the reflected wave, helping create more torque with no penalty in top-end power.

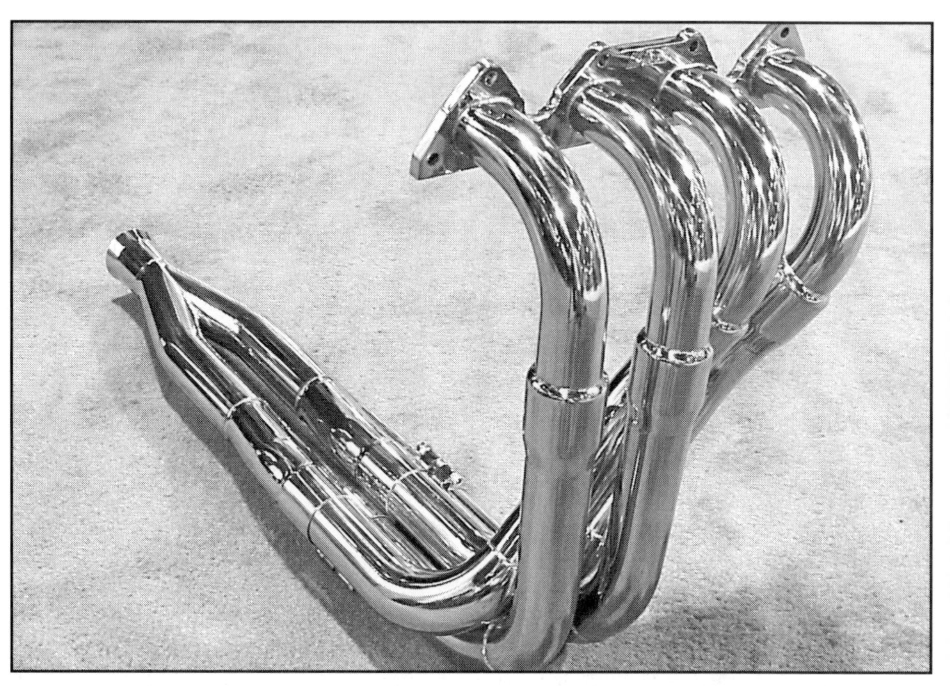

High Tech Exhaust also makes a street version of their racing anti-reversion header.

headers are the current craze in the rapidly growing All Motor drag classes. These cool machines are running in the 11- and 10-second range, speeds that were once only touched by the quick class turbocharged cars just two short years ago.

The area where an anti-reversion header differs from a regular header is in the primary tubes. An anti-reversion header has a stepped, primary tube. The is an area a few inches away from the head flange where the primary tube will suddenly step up in diameter, going from 1.5" to 1.75" in diameter or from 1.75" to even 2.00" on some of the racing versions. In some cases, a larger diameter reversion trap chamber is used instead of a stepped primary tube. This looks like a goiter on the primary tube. The theory of the stepped diameter is that the step will help prevent back flow or reversion of the exhaust gas during the overlap period, when both the intake and exhaust valves are open.

The theory is that if the sonic-reflected sound wave comes up the primary pipe slightly early or late relative to engine rpm, it won't not match the pipes' "tuned" length, and its arrival at the exhaust valve may correspond to the valves being on overlap. If this is the case, then the cylinder can be contaminated by exhaust gas pushed into the cylinders by the reflected wave. The tube step in the primary pipe or the larger diameter anti-reversion chamber will help damp out this reflected wave at rpm other than what the pipes' length was tuned for, effectively making the headers tuned period longer. This creates a wider powerband and more torque.

Basic Design—The design of the anti-reversion header is more art than science as there are no mathematical models for the proper way to calculate the proper dimensions for the anti-reversion effect to work at its best.

Comptech and Mugen make prime examples of a stepped primary anti-reversion header. Hi-Tech Exhaust is noted for their extensive pioneering use of stepped primary diameters and anti-reversion chambers in their header design with extensive use of merged collectors. In fact a Hi-Tech Exhaust header is one of the secret weapons responsible for the incredible quarter-mile times that the latest imports record. I have seen a High Tech Header make 18 more wheel hp over a DC fabrication header on a highly modified B18C motor.

THE INTAKE SYSTEM

The intake system's job is to take outside air, clean it, and bring it to the engine so it can be mixed with fuel and burned. Sounds like a simple job but there is usually power lurking here for the weekend tuner to unleash. As we stated before, the engine is mostly a giant air pump. The easier we can make it for air to be sucked in, the less power the engine will have to waste to get the air inside to where it can be used.

Drop-In Air Filters

For many, a first step is to replace the sometimes-restrictive factory

Although this air intake does pick up cold air, its design does not work too well as a ram air setup like its strange position was probably intended to. Jousting anyone!

The good old K&N cone filter is used on many cold-air intakes or sometimes just bolted to the stock air inlet in place of the air box.

AEM's cold-air intakes are dyno proven to make plenty of power. In testing by *Sport Compact Car* magazine, the AEM cold air intake out powered the competition.

paper filter element with a high-flowing one. K&N makes OEM replacement, high-flow air filters for just about every car made on earth. The K&N filter uses a free-breathing, oil-saturated gauze filter that can flow up to 100% more that the stock paper element. The K&N filter is also very durable and can be washed and reused many times over. This makes a K&N a good money-saving addition to a car. Usually a drop in-filter gives from 0-2 more hp on a Honda motor.

A limitation of the drop-in filter is that you are restricted in filter area to whatever the vehicle's manufacturer originally designed. If the car was designed with a tiny air filter, your high-flow drop-in will be tiny also. If the car has a restrictive airbox, you are stuck with that using a drop in.

Cone-Style Open Air Filters

In the search to make modern cars more and more quiet, some manufacturers have been adding silencers to the air intake of their air boxes. This can sometimes make the air intake quite restrictive. As most of us enthusiast drivers feel that intake noise is music to our ears, we do not mind adding more of it to our car's mechanical symphony. In that case, there are many companies that make cone-type air filters. These filters are racing-type cone-shaped air filter elements that have a large area for filtering. The more surface area, the less restriction to the incoming airflow a filter will have.

Cone filters typically have an adapter that bolts directly to the car's intake pipe or airflow meter, replacing the stock air filter and air box. The best adapters have a radiused inlet to help smooth the airflow into the intake tube. These cone filters typically make from 2–3 more hp on a Honda and are much more noisy than stock. A K&N Filtercharger Kit, the HKS Powerflow and the Greddy Airinx, are examples of a cone-type open filter intake. The HKS and the Greddy feature open-cell foam filter elements and are also very free flowing.

Air Intake Systems

Finally there are complete air intake systems, ram air systems and cold air intake systems. Most air intake systems use a long mandrel-bent pipe to replace the restrictive stock, convoluted, rubber pipe. The best of these systems consider the tuned length of the pipe and take advantage of the incoming pulses to get a ram effect, kind of like headers but in reverse. Most of these intakes range from 2.25" to 3" in diameter with lengths from 12–30 inches in length. The DC Sports DAC, RS Akimoto Funnel Ram, and Weapon R are all examples of these types of ram air intakes.

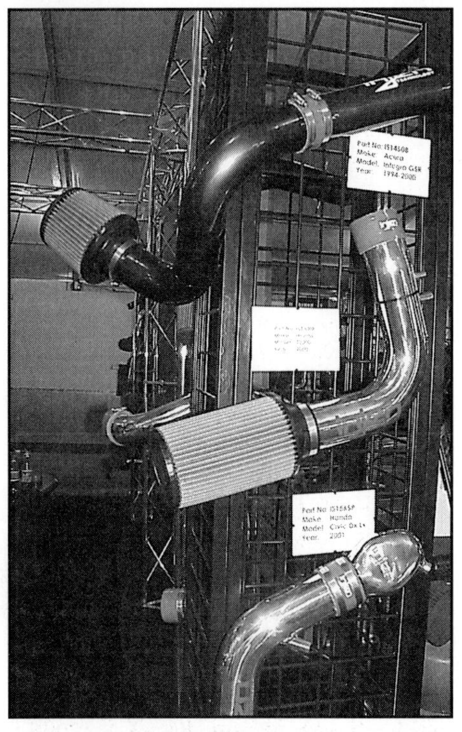

Injen makes a wide assortment of air intake for every Honda application.

DC Sports cold air intake system is beautifully made of polished stainless steel.

This beautiful and expensive carbon fiber ram air system for the S2000 is made by Mugen and is one of the few air intakes known to actually make power on this well-engineered car.

Some of these systems can be plumbed to the front of the car with a scope to take advantage of high-pressure air bombarding the front of the car to help ram more air into the engine. A true effective ram air system can give a 1–3% power boost over a conventional tuned air intake. The Iceman competition air intake takes the place of one of the front headlights, replacing it with a scoop for a very direct ram air setup. Air intake systems can be plumbed outside of the engine compartment to take advantage of cold air, unheated from the engine, radiator or A/C condenser. A rule of thumb is that for every 10 degrees you can make an engine's intake air cooler, you can get 1% more horsepower. The colder the intake air, the denser it is, and the more oxygen it contains. This is where the additional power comes from. These systems can produce from 4 to 6 more horsepower at the rear wheels depending on how they were designed and what speed the power was measured at (in the case of ram air). AEM makes the most well known of the cold air intakes. *Sport Compact Car* tested Civic air intakes and found the AEM intake to produce the most power.

A properly designed system can use the reduction of restriction, tuned length, ram air and cold air to get tremendous increases in power. A good free-flowing air intake can also increase gas mileage by reducing pumping losses.

On the downside, many cold air and ram air set-ups can also scoop up water during a rainstorm, which can possibly damage your engine. Look at

AEM's innovative anti-suck valve prevents damaging water ingestion from a low hanging cold-air intake system. In this dramatic demonstration, a completely submerged filter did not pick up one bit of water when sucked by a huge industrial vacuum cleaner. In testing done by *Sport Compact Car* magazine, an NSX run at full throttle on a dyno would not suck up any water, even if the filter was completely submerged in a bucket of water! AEM's valve is a great safety device.

Here, an Integra Type R manifold is being Extrude Honed. Extrude Honing ports and polishes imposible to reach passages inside a manifold by running an abrasive filled putty through them at high pressures.

The JUN manifold is as good as it gets: a large plenum of the correct volume for maximum power, an irregular, non-resonant shape, and velocity stacks for the runners inside the plenum.

the designs carefully and be careful if they can pick up water. Many cold air intakes, such as AEM, have a removable section so the air filter can quickly be mounted higher in case of rain, which is a definite plus. Best of all, the AEM cold-air intake has a special bypass valve that will automatically divert the airflow to a higher intake if the cold air part of the intake sucks up water for any reason.

As we mentioned before, air intakes and cone-type filters are also much louder than the stock airbox. One of the reasons why the stock air box is restrictive is because it contains baffles to help silence the roar of the air entering the motor, which can be louder than the note of the exhaust. Fortunately this roar can be mostly limited to wide-open throttle but it is an important point to consider for those who are sensitive to noise.

The Intake Manifold

There are a couple of intake manifold options for B-series motors. A popular one is to fit the IntegraType R manifold, which is a direct bolt on for B16A motors. The Type R manifold has shorter, larger diameter runners. The manifold is worth about 4-6 more hp from the midrange on up with a slight hp penalty at lower rpm. Skunk 2 makes a copy of the Type R manifold for the B18C. This gets similar gains to the Type R manifold on the B16A. Prelude owners can install the JDM Prelude Type S manifold on their motor. The Type S manifold is good for about 2–4 hp.

JG/Edelbrock, STL, Eshleman, JUN and Venom make a short-runner, large Plenum manifold for the B16 and B18C, which helps power in highly modified motors. This manifold has produced upwards of 30 hp on highly modified turbo motors. However, for mild street motors it is probably best to stick with the stock or Type R manifold. The large plenum of these race type manifolds can make throttle response slightly sluggish. They will work nicely on turbo motors and fairly built naturally aspirated motors. The best of these manifolds have a plenum volume of about 1.5 times the engine displacement, have non parallel walls so they will not resonate

The STL manifold is used by fast racers like Lisa Kubo. It has nice radiused inlets to the intake runners and a good-sized plenum. Its boxy shape has parallel walls which compromises tuning somewhat.

The Venom manifold is very similar to the STL.

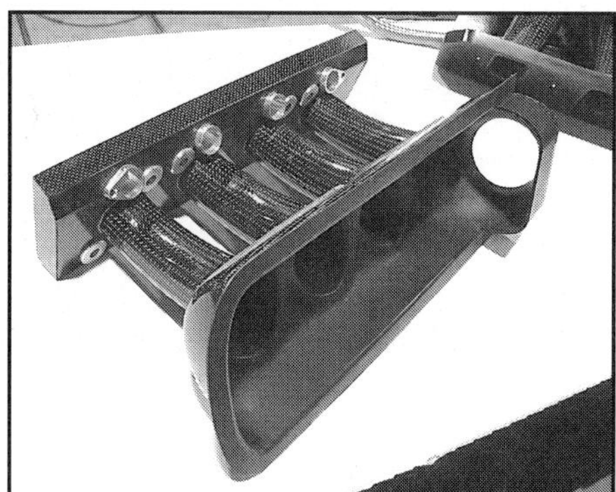

Eshelman's carbon fiber manifold is another optimized design like the JUN manifold. Its carbon construction is very light, like a shoe box! The cost, don't ask!

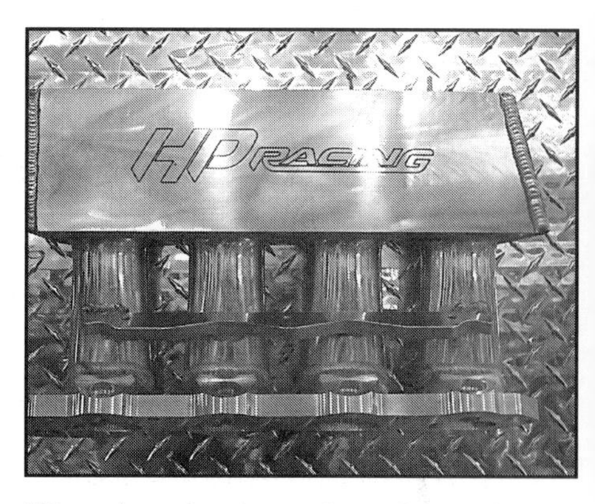

HP racing also has a box plenum intake manifold.

TWM's manifold with dual side-draft throttle bodies, although becoming popular for all motor class drag racing, also works well on the street and for road racing due to its wide powerband and sharp throttle response, the only drawback is the cost which includes a stand alone EFI system.

and cause lumps in the powerband and have radiused entrance velocity stacks built into the ends of the runners.

A combination that is only now coming into wider use, especially in the all-motor classes is a short individual runner manifold with twin two-barrel sidedraft throttle bodies. Although this setup looks pretty race only, it can be used on a totally streetable car. The small runner volume and one throttle valve per cylinder give very sharp throttle response for a fairly decent bottom end while the short runners give good top end also. TWM makes Honda applications for both the manifold and throttle bodies. The only real disadvantage is that this setup must be used with a user programmable stand-alone fuel injection system which can

Here is another TWM setup although this one has a plenum chamber for a turbocharged application. This is Gary Gardella's winning Quick Class racer.

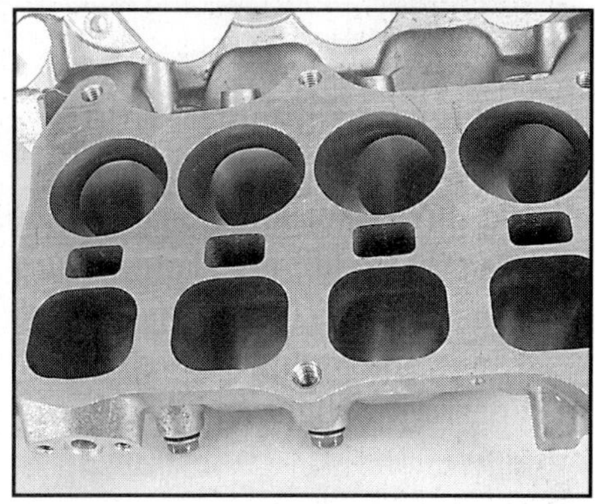

The Extrude Hone process ports and polishes even the most intricate intake manifolds where no human can reach. Look at the smooth insides of this B18C manifold.

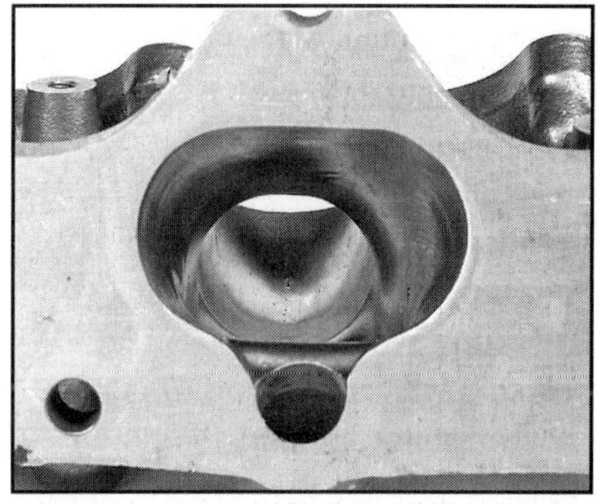

Here is a look up the runner of the same manifold after Extrude Honing. Note how the runner is ported and polished farther up the runner than anyone could possible reach with any handheld tool.

The JG/Edlebrock throttle body is a big 65mm.

Eric's Racing makes a CNC machined billet throttle body, which can be made as big as 75mm.

make it a costly proposal.

For mild street motors, or motors where there is no other manifold option, the Extrude Hone process has been found to be beneficial. Extrude Honing is the process of forcing an abrasive paste through a manifold at high pressures, to port and polish the internal parts of the manifold where a porters bit cannot reach. Extrude honing the stock intake manifold has also been found to produce about 5 more hp. A Type R or Type S manifold treated to extrude honing is an excellent piece for street use.

Big Bore Throttle Bodies

Boring the throttle body has been found to sometimes increase power on Honda engines. The D-series Civic motors can usually benefit the most from a larger bore throttle. The benefit of a bigger throttle body is quicker throttle response and a little more power. On an engine with all the bolt ons, power gains from 1–4 hp have been recorded. RC engineering and JG supply throttle bodies that are bored out and fitted with about a 4mm larger throttle plate. Because of the improvement in throttle response, a bored out throttle body feels like more than this amount.

UNDERDRIVE PULLEYS

Underdrive pulleys are an often-

The Unorthodox Racing UD pulley set shaves several pound of rotating weight off of the motors reciprocating parts, allowing the engine to rev freely as well as freeing up an additional six dyno-proven horsepower to the wheels.

The AEM power pulley set, does not change the crank pulley but uses different size accessory drive pulleys to reduce the drive ratio and free up power to the wheels.

overlooked performance part that can give you a pretty large bang for the buck on your Honda. An underdrive pulley set does exactly like its name suggests, it drives all of the accessory drives on your motor like your power steering, alternator, air conditioning and water pump about 20 percent slower than stock. Incredibly, this simple modification gives about 4–6 more hp to the wheels, more than some brands of headers as an example!

Although underdrive pulleys reduce the parasitic drag of the engines front-end accessories, they do reduce the effectiveness of them by slowing the drive speed. Although we have not experienced any problems with charging, steering effort or engine cooling, you must be aware that the A/C will work a bit less well at idle and if you have a big thumping stereo, the alternator may not charge the battery quickly enough if you are pounding the bass.

Unorthodox Racing makes underdrive pulleys for all popular Honda applications. The Unorthodox pulleys replace the stock heavy steel pulleys with CNC aluminum billet machined lightweight pulleys with an approximately 20% lower drive ratio. The Unorthodox pulleys eliminate the stock heavy steel front pulley/harmonic balancer with a lightweight solid aluminum front pulley that both eliminates rotating weight and reduces the accessory drive ratio. Since Honda engines are internally balanced, the harmonic balancer is not as critical for engine longevity as an externally balanced American V-8 and we have not heard of any problems with Unorthodox pulleys even in long-term racing use.

AEM also produces a pulley set that does all of the drive ratio reduction at the accessory instead of relying on a small crank pulley like Unorthodox does. AEM feels that it is safer to retain the factory harmonic balancer/front pulley for best engine life. AEM claims that the extra mass of the stock pulley helps damp out potential harmful engine vibrations. Only a few years of particular Honda motors actually had a mass damping front pulley with a rubber mounted inertia ring and since all Hondas are internally balanced, we don't think the low mass Unorthodox pulleys are any problem. A drawback to the AEM setup is that it does not eliminate about 3 lbs of rotating weight from the crank like the Unorthodox pulleys will. We feel that both pulley sets are safe and proven and both produce similar increases of power at the wheels. We have never had any engine trouble that was directly attributable to them. H-series Prelude motors even have balance shafts so pulley issues should not affect these motors.

We like to get something for nothing and underdrive Pulleys are almost that, getting a good increase of power for almost no penalty.

DYNO TUNING

Dyno tuning can offer a great return on your investment, and when done correctly, will extract every last bit of power from your modifications. Proper tuning involves buying dyno time, which can be expensive, an hour or two on the dyno is a lot cheaper than buying some widget and usually yields more power than any single typical trick part. I feel that dyno tuning is so important that the subject belongs in this chapter on basics, rather than on advanced modifications.

The Dynojet

Just a few short years ago, dyno tuning was usually only done by race teams on a big budget who could afford to run their engines on an engine dyno. To do so is usually very expensive, because the engine is run separately from the car, requiring special mounting hardware, electrical hookups and the particular ECU of your engine. Such time-consuming and expensive factors made dyno tuning a luxury that the usual enthusiast or amateur racer could not afford to do.

Enter the Dynojet. The Dynojet chassis dyno was introduced several years ago as a low-cost way to dyno test high horsepower cars. On a Dynojet, a computer measures the amount of time the engine takes to spin a large, heavy (4000–8000 lb.) roller up to speed throughout a gear range (usually 3rd or 4th) from near idle to the redline. Unlike other chassis dynos, the dynamic nature of how the Dynojet applies load reduces the chances of wheelspin. The lack of wheelspin makes Dynojet readings very easy to repeat for consistent readings. It is fairly common to achieve repeated runs that only vary within one horsepower when run under the same conditions.

Buying Dynojet time is fairly cheap to purchase because the overhead is much less expensive to operate one. Dynojet facilities do not require a lot of special building modifications such as a concrete lined pit, extra electrical power or high pressure water lines and noise limiters and special EPA exhaust/smog controls. Because of its relatively low cost, the Dynojet has been a very popular addition for many local speed shops.Most shops charge about $100 per hour if you do all of your own wrenching. While that may sound like a lot, I can just about guarantee you'll get your money's worth in "found" horsepower. Here are some basic dyno tuning tips to get the most out of your dyno tuning session:

1. Have a plan on what you are going to adjust and learn how to do it quickly before you get to the shop. When the clock is ticking, you don't want to be figuring out how to adjust your ignition timing or fuel pressure. Maybe even practice, doing a dry run before you head out.

2. Bring your own tools, and make sure they are the tools you'll need. Most shops will not loan tools to customers.

3. Read this guide first before you start making adjustments to increase your chances for success. Many of the tips in this book will help you avoid simple mistakes.

4. Change or adjust only one thing at a time. Take exact notes of your changes and the results.

5. Many shops won't let customers work on their own car while it is on the dyno. To make the most of your time, write out a plan of the adjustments you intend to make, and show it to the dyno technician before you begin. In this case, prepare to pay a little more. Another alternative is to go to a dyno shop that has a good local reputation for their tuning ability and just hand your car over.

6. Most shops will make you sign a waiver saying that they are not responsible if your car blows up. But, you'd have to do something drastically wrong for this to happen. Chassis dyno tuning is fairly safe.

7. To find a shop with a Dynojet go to www.dynojet.com for the location of the Dynojet closest to you.

Prepping the Car

But before making test runs, you must prepare the car. First, make sure you set your tire pressures and record them. You want to run the same tire pressure every time you go to the dyno. It is also important to run on the same wheels and tires each time as wheel weight and tire rolling resistance makes a difference from run to run.

If you are raising or lowering the car it is important to have your alignment set the same way each time also as changes in ride height can affect alignment and add power-robbing tire scrub. Changes in ride height also affect CV joint angularity and the amount of power they absorb.

Once the car is strapped to the Dynojet, you may want to have the operator do a "coast down" test. This measures how much power is lost due to minor variation in the car's angle on the dyno. Any difference in coast-down power loss can be added or subtracted from previous dyno runs to go the final bit to ensure maximum accuracy.

During the initial runs, the power figures should not be considered for

AEM adjustable timing gears are an easy-to-use, no-slip, high-quality design. The AEM gears are available for all popular Honda motors.

tuning adjustments until horsepower levels have stabilized. The engine must be fully warmed up before you can trust the results. Most naturally aspirated cars make the most power on their second run, and most will achieve stabilized power results after 3 or 4 runs. On naturally aspirated cars, it takes this due to the heating of the intake tract. Turbo and supercharged cars usually make their best power on their first pull before the intercoolers and intake tract become too hot. The variation between a cold run vs. a hot run can be as much as 5 hp in a naturally aspirated car or an amazing 30 hp in an intercooled turbo car running lots of boost!

In addition to warming the engine, the gearbox must be warm as well, and the best way to do that is to drive a minimum of 15 minutes on the freeway or around town to ensure that the gearbox oil is up to operating temperature. A cold gearbox with thick oil can rob as much as 3–5 hp.

Run the dyno tests at the same "smoothing factor" each time. The smoothing factor is an averaging equation that the Dynojet software can run to "smooth out" the dyno chart. The higher the smoothing factor, the lower the peak horsepower will be, and the smoother the horsepower trace on the dyno chart. For the best back-to-back comparison, the smoothing factor must be set to the same number.

A lower smoothing factor car shows a higher power peak but that is not a realistic way to judge increases of power. A smoothing factor of 3 is usually good for tuning.

Run the car with the hood up, and if possible, direct a small fan at the air intake to prevent hot air from the radiator fans from affecting the test results. I have seen variations of as much as 6–9 hp on a Civic Si from the radiator fan clicking on and blowing hot air into the intake during a test.

Finally, make sure the dyno operator diligently enters the correct humidity and that the dyno's barometer and temperature sensors are working correctly. That being done, you are ready to do some adjustments and make some horsepower!

Adjustments

Once the car is warmed up and you are satisfied that the car's output has stabilized, you can begin playing around with various settings and make some adjustments. We will list adjustments in the order they should be performed to achieve the most efficient tuning.

Ignition Timing—The ECU or engine control unit in your car is usually programmed on the conservative side. It has an ignition curve programmed so if some cheap sap fills the tank with low-octane gas, the car won't detonate and fry your pistons. But if you are a hardcore performance nut and are willing to shell out the extra bucks for 92 octane or better, you can fiddle with your ignition timing to extract a bit more power.

Once the car is strapped to the Dynojet and stabilized, try advancing the timing two degrees at a time and observe if the power increases. On most cars it will. If you advance timing more than 4–5 degrees above the factory spec, use extreme caution. Do not tolerate any detonation at all. Detonation can damage your engine big time, so don't let it go on for more than a second or two. If you hear the telltale ping or metallic ring, back off the gas at once.

Refer to your factory service manual for the exact instructions on how to adjust the timing. On Hondas, loosening the bolts that hold the distributor in place and turning the distributor do this. Don't just guess and twist, use a timing light and do it right.

Don't assume that if some advance is good, more is better. Advance the timing in two degree steps until the power either flattens out or decreases, then back down slightly from there. Later, when road testing your car, if you note any detonation, you may have to reduce your timing even further, because real road conditions load the car a little differently than the dyno. Typically, you might pick up 2–5 hp by experimenting with better fuel and advancing your timing.

Cam Timing—Hondas, like most cars, have their cam timing set at the factory for minimum emissions with the production of maximum power a secondary goal. As we explained earlier in this book, most production camshafts are designed for minimum overlap with a wide-spread lobe separation angle to reduce hydrocarbon emissions and to ensure a smooth and stable idle. Since the enthusiast does not worry about these factors like the OEM engineer, power is usually available to those who are willing to spend a little time with tuning. Adding little bolt-ons like air intakes and headers can also change the optimal cam timing that your car prefers. V-Tec equipped Honda and Acuras with their big duration secondary lobes can almost always benefit from some cam timing adjustment. Even the D-series SOHC motors usually respond well to a little tweeking in this department.

An interesting fact is that almost all aftermarket and many stock cams are usually ground a degree or two off from the spec that they are supposed to be ground to which is normal variation of tolerances in production. Dialing in the cam timing allows the tuner to compensate for these slight errors. Often there are a few hp just waiting to be unleashed by this alone.

To dial in the cam timing, adjustable timing gears are needed. The market is ripe with adjustable timing gears for Hondas and many manufactures have various adjustable gears for sale. As a note, some poorly designed adjustable gears can slip, throwing the cam timing off. This can cause anything from a loss of power to total valvetrain failure if the cam timing gets so off that pistons hit valves.

With the use of a chassis dyno, like a dynojet, the accepted way to dial in a cam is to start with the intake cam. Do not try to tune cams by seat-of-the-pants feel. It is easy to get fooled by this. Since most street-type engines respond well to having the intake cam advanced, on DOHC engines like the B-series and the H-series motors, advance the intake cam 2 degrees at a time, testing in-between adjustments until the power falls off or the desired powerband is met. If the power falls off right from the get go, try retarding the intake cam although this is not likely on most street type engines. After adjusting the intake cam, most tuners first retard the exhaust cam slightly. Try retarding the exhaust cam in one to two degree increments until the power drops. If the motor does not respond to this try advancing the exhaust cam. Many motors seem to like having the intake advanced 4–5 degrees and the exhaust retarded 1–2 degrees. My take on this is that doing this increases overlap, which done in moderation tends to broaden an engine's powerband. This also helps production type motors with their more restrictive ports, intakes and exhausts (as opposed to real race specific designed racing motors) breathe better at mid and high-mid rpm. As a warning, this is very general and must be verified on a dyno as many motors will not like this either!

While dyno tuning, if you progress in the adjustment orders listed below, you can usually get results with a fewer number of dyno pulls than proceeding willy-nilly. You don't necessarily have to go through the whole sequence of adjustments when tuning; if a series of adjustments starts to make good results, continue in that direction, do not go through the whole sequence. For instance, if advancing the intake cam moves the powerband to where you want it, you don't need to try retarding the cam also.

When dyno tuning on a SOHC motor, you should experiment in this order, advance cam, retard cam. Two-degree increments usually make a big enough change to easily spot on the dyno. For a DOHC motor, work in this order in two degree increments: advance intake cam, retard exhaust cam, advance intake only, retard intake only, advance both intake and exhaust cams, retard both intake and exhaust cams, advance exhaust cam, retard exhaust cam.

As a warning, piston-to-valve clearances decrease when advancing the intake cam or the intake lobes in the case of a SOHC engine. Usually this is not an issue in stock engines but with a higher lift and long duration racing cam and milled heads or other modifications, it is possible to hit the intake valves on the piston and cause damage. In most cases you can safely advance the cam on a street engine at least 6 degrees before you need to worry about contact, however, if in doubt, measure the clearances before proceeding.

Here is a rough guideline on how quick and dirty cam adjustments can affect your motors powerband.

Advance both intake and exhaust cam

Increases bottom-end power while decreasing top-end power, closes the exhaust valve earlier and opens the intake sooner to move the whole overlap period earlier where it works to scavenge cylinder better at low rpm. For SOHC engines this is one of your only two choices for cam timing changes. If you go too crazy with this adjustment you stand a chance of hitting intake valves to pistons. Can increase the chance of detonation at low RPM.

Retard both intake and exhaust cams

Increases top-end power while decreasing bottom end. The opposite of advancing the cams, it moves the overlap period later in the cycle to where it is more effective for scavenging the cylinders at high rpm. For SOHC engines this is your other only choice for cam adjustments.

Advance intake only

Increases overlap and starts overlap period earlier. This helps bottom-end and midrange power, usually without sacrificing too much top-end power. Works best usually on stock head or heads that flow poorly. Usually engines that have long strokes and short rods like this. This will usually give you more lope to the idle. This is a good first thing to try when optimizing cam timing on a dyno for a street car with mild cams.

Retard intake cam only

Starts overlap period later, reduces overlap. Usually helps top-end power at the expense of lower mid-range especially on big duration cams in race-prepped motors with good port flow and favorable undersquare bore stroke ratio and stroke to rod length ratios over 1.7:1. Usually stock type street motors don't like this too much. Not one of the first things to try. Smooths idle.

Advance exhaust cam only

Reduces overlap and increases blowdown by opening the exhaust valve sooner. Sometimes this is not good because this is shortening the power stroke. Usually helps top end power at the expense of lower mid range especially on big duration cams in race prepped motors with good port flow and favorable undersquare bore stroke ratio and stroke to rod length ratios over 1.7:1. Usually stock type street motors don't like this too much. Smooths idle. This is not one of the first things to try.

Retard exhaust cam only

This increases overlap and helps bottom midrange. Works best usually on stock head or heads that flow poorly. Usually engines that have long strokes and short rods like this. This will usually give you more lope to the idle. This is a good second thing to try when optimizing cam timing on a dyno for a streetcar with mild cams. Usually most cars do not like more than a couple of exhaust cam degrees of retard.

When adjusting cam timing on the dyno, remember it is better to go for area under the power curve rather than maximum peak hp. Having 4 more horsepower from 5500 to 7800 rpm will get you down the track faster than having 7 more from 7500 to 8000 rpm with less everywhere else. Adjust the cam timing to get the most power in the rpm range where the car will be pulling the longest in each gear.

As a warning, piston-to-valve clearances decrease when advancing the intake cam or the intake lobes in the case of a SOHC engine. Usually this is not an issue in stock engines but with a higher lift and long duration racing cam with milled heads or other modifications, it is possible to hit the intake valves on the piston and cause damage. In most cases you can safely advance the cam on a street engine at least 6 degrees before you need to worry about contact, however, if you are not sure, measure the clearances before proceeding. A mistake here can be costly.

It can be possible to see gains of up to and over 10 hp by optimizing cam timing although 3-5 hp across the rpm range is usually more typical. It is also important to remember that any change to the motor, especially to the intake and exhaust system can change the cam timing that the motor likes. It is also important to know that the motor may like slightly different ignition timing after the cam timing has been set so it might be a good idea to play with that a little once after optimizing the cam timing.

For turbocharged motors, it usually helps to reduce overlap somewhat by first advancing the exhaust cam and possibly retarding the intake cam, especially with smaller, high backpressure turbochargers. Turbocharger exhaust manifold backpressure should be verified to help predict how the engine will respond to tuning.

The same goes for supercharged motors with bigger-than-stock cams; slightly advancing the exhaust cam first may help.

Large, race-type turbo applications may respond to tuning pretty close to what a naturally aspirated motor likes, just with a slightly more spread lobe center, usually with 2-8 degrees less overlap.

FUEL PRESSURE

Due to the fact that auto manufacturers have no control over how the end user is going to drive and what octane gas will be used, the wide-open throttle air/fuel ratio programmed into the car's ECU tends to be set on the rich side for most cars. A richer air/fuel ratio burns cooler and is safer by being more resistant to detonation. Modern engines are programmed to run a tick richer than stoicmetric or a 14.7:1 ratio (a perfectly chemically balanced) of air to fuel at low speeds and light throttle because this is where a catalytic converter works with the greatest efficiency. At wide-open throttle the best power is usually obtained at a richer mixture ratio of about 12.5–13:1.

For an extra margin of safety, most modern cars run a more conservative 11.5–12:1 richer mixture at wide open throttle to aid cylinder cooling and reduce the chance of detonation with low-grade fuel. Most factory turbocharged motors run as rich as 10.5:1 at wide-open throttle.

To get more power, if you intend on running only premium fuel, it will respond to being leaned out a little at wide-open throttle

Hondas use what is called a Speed Density airflow metering system, which uses a MAP or Manifold Absolute Pressure Sensor to help the engine's ECU to determine how much fuel to inject for a correct mixture.

AEM's pressure regulator is an accurate, high quality unit.

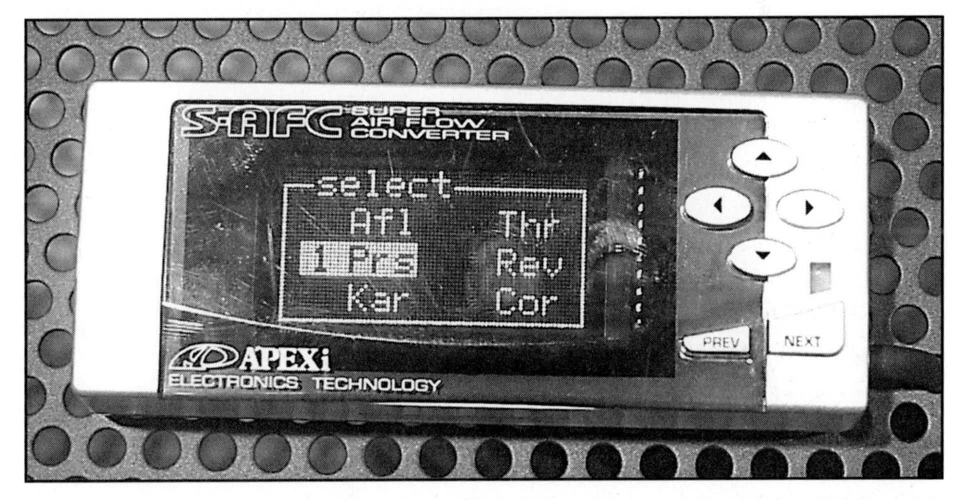

The Apexi S-AFC controls not only the fine tuning of the mixture, but the VTEC activation point also.

This sensor reads manifold pressure or vacuum, which, together with data from the throttle position sensor, gives the ECU an idea of how much fuel to inject into the engine. Unfortunately, speed density systems like Hondas do not compensate for modifications very well which can cause irregular air/fuel ratios if camshafts, air intakes and other goodies are added to the motor.

Fuel Regulators

All is not lost, however. You can add a device called a fuel pressure riser. This is a vacuum referenced adjustable pressure regulator that goes in the return line to the gas tank from the injectors. A fuel pressure riser has a manifold vacuum reference port that goes to the intake manifold. By using this reference, the riser can richen the fuel mixture mostly at WOT. With a fuel pressure riser, you can control the fuel pressure to the injectors allowing you to increase the pressure to add more fuel to make up for the additional air being drawn into the engine. Stillen and AEM make high quality fuel pressure riser kits. Hondas using ram air systems like the Iceman race setup almost certainly need to run more fuel pressure to avoid leaning out at high speeds. Tuning these is best done at the track. Turn up the fuel pressure 5 psi at a time until the trap speed falls, then back up to the previous best trap speed.

Another way to fiddle the mixture a little is with an electronic fuel device like an Apexi AFC. This alters the MAP signal to fool the ECU into injecting more or less fuel. Apexi also has an SAFC for V-TEC equipped cars; the SAFC controls the V-Tec activation point as well as fine-tuning the fuel mixture. Activating the secondary lobe of the camshafts earlier has proven to give more midrange power if the correct fuel mixture can be supplied at this point.

Optimizing the fuel and air mixture for more power will increase cylinder pressure. This may cause detonation necessitating a reduction in timing. In other words, you may have to do some fiddling between optimal mixture and timing for best power but with no detonation detected optimize your set-up for both. As a rule of thumb, if you are suffering from borderline detonation at your optimal power tuning point, it usually hurts power less to richen the mixture up to suppress detonation rather than reduce timing so try adding fuel first before taking out spark.

The best thing to do is to first try slightly leaning out your mixture on the dyno in 5% increments on your fuel controller or by trimming your fuel pressure by 5 psi or so at a time. If the car makes more power, go a little more at a time until the power gain flattens out, then richen a little to right before the flattening point. Be extremely careful, do not tolerate any detonation and use only premium fuel. It is not a good idea to go much past 10% leaner. If going leaner does not help try richer using the same methodology. Going richer is much safer as there is less chance of detonation by a richer mixture. If you are increasing fuel pressure to richen the mixture, it is very dangerous to exceed 80 psi of fuel pressure because this is the point where many injectors can stick, due to too much side load on the injector pintile. This can cause erratic running and a dangerous overly lean mixture. 80 psi is very extreme and you should never be turning the fuel pressure up this high on a naturally aspirated car, but it is possible for a bolt-on supercharger or turbo kit with a mis-adjusted rising rate fuel pressure regulator to run this high.

It is possible to get from 0–10 hp tuning the fuel mixture on bolt-on equipped streetcars.

PLUG READING

An important skill for tuning, especially when it comes to judging the correct amount of timing advance and how lean the mixture is the lost art of reading plugs. Reading plugs is looking at the porcelain insulator of the center electrode and the general condition of the center electrode and ground electrode. To read plugs, it is handy to have a magnifying glass and a light or better yet, a lighted magnifying glass. When the car completes a dyno pull, the engine should be quickly shut off and the plugs inspected without having the car idle for a long period of time.

The first things to look for on plugs are signs of detonation. The usual signs of mild detonation are dark gray or silvery specks that contrast against the light colored porcelain of the center electrode, much like pepper on a white dinner plate. The specks are about the same size or slightly smaller than pepper also. Unfortunately the specks are caused by aluminum being vaporized off of the tops of your piston or the combustion chamber by the heat and pressure of detonation! If you look at the ground and center electrode, you may see, on a detonating motor, tiny dingle balls that look like spherical weld splatter except very small, so small that it may only be possible to spot them with the magnifying glass. The ground and center electrode may have a blue-green to straw-yellow tinge on it also.

If you see these clues, you may not have heard the detonation over the roar of the motor on the dyno. You need to either add mixture, back off the timing or a combination of both, whichever causes the least loss of power, or serious damage to the engine will result. Broken ring lands, burnt valves or sometimes even holed pistons and pounded rod bearings result from detonation.

In more dire cases of detonation, the ground electrode may be completely or partially melted away. In even worse cases, the ground electrode may be damaged or destroyed. In very severe detonation, the porcelain center electrode insulator may be fractured and partially missing from the hammering pressure waves of detonation. If you observe damage to the plugs like this, it is prudent to do a compression and leakdown test before proceeding further as you have probably seriously wounded your motor.

You can also determine if your

If a spark plug has too cold a heat range, it will foul. If it's too hot, the plug may self-destruct. This plug is an example of the latter. It got so hot that the center electrode was completely burned away. Photo by Dave Emanuel.

engine is running too rich or too lean by looking at the plugs. With modern unleaded pump gas, proper mixture nowadays will have a center electrode porcelain color anywhere from bone white to medium brown. In the old days due to the large amounts of lead found in gasoline, the trick was to shoot for a chocolate brown. This would be terribly rich using modern fuel. Pump gas usually colors to about bone white to light tan and racing fuels are from light tan to medium brown. Modern racing fuels increase octane mostly by the addition of heavy, long-chain aromatic hydro-carbons rather than adding high concentrations of poisonous, plug-coloring tetraethyl lead like in the old days.

If an engine is running way too lean, the center electrode will be bleached white with kind of a flaky, peeling look, like it was sandblasted with really coarse sand. Usually a plug showing this lean will also show signs of detonation. Too rich will be a dark brown or black plug covered with dry soot. Serious rich will be black and wet smelling strongly of gas. Fuel-injected cars seldom look like this unless there is a malfunction like a stuck injector or super high fuel pressure sticking an injector pintle. A worn motor burning oil will have the plugs covered with a sticky black carbon.

Heat Range

A plug's heat range is something to consider also. The heat range is a rating of the plug's ability to conduct heat away from the center electrode. A stock plug is usually a hot heat range, so the plug does not foul up while spending a lot of time idling and driving at low speeds. Racing plugs are cold plugs so the center electrode will not get too hot and contribute to detonation. However, racing plugs will quickly foul with carbon deposits driving at low street speeds. For a bolt on modified street motor you can figure out the part number coding on the type of plugs your car takes and install one heat range colder if you are running at high speed for an extended amount of time like on a road course. If your motor has higher compression or has a turbo, blower or NOS added you should automatically go up at least one heat range colder and possibly two if you are doing some track driving. If you are only racing a heavily modified car with no street driving, several heat range colder racing plugs may be needed.

Physically, the difference between a hot plug and a cold plug is the length of the insulator for the center electrode. A hot plug will have a longer porcelain insulator for the center electrode that extends deep within the steel shell of the spark plug. This creates a longer path for heat to be conducted out of the center electrode so the electrode will operate at a higher temperature. A cold plug will have a center electrode insulator that is much shorter internally so there is a much shorter path for heat to be conducted away from the center electrode ensuring that it will run colder.

So hopefully you have learned some of the basics to tune your motor. Like we said before, investing a few hundred dollars in dyno time will usually get you more power per buck in a mildly built car than any similar value of shiny, expensive go-fast parts.

So go ahead, buy those parts, learn how to make the adjustments and go and book some time on your local Dynojet. If you don't feel confident, perhaps a local shop with a Dynojet and experience in tuning Hondas can help you out.

CAMSHAFTS

The next subject in our quest for speed and power are the camshafts. These are the bumpy sticks that reside in your cylinder head that spin at one

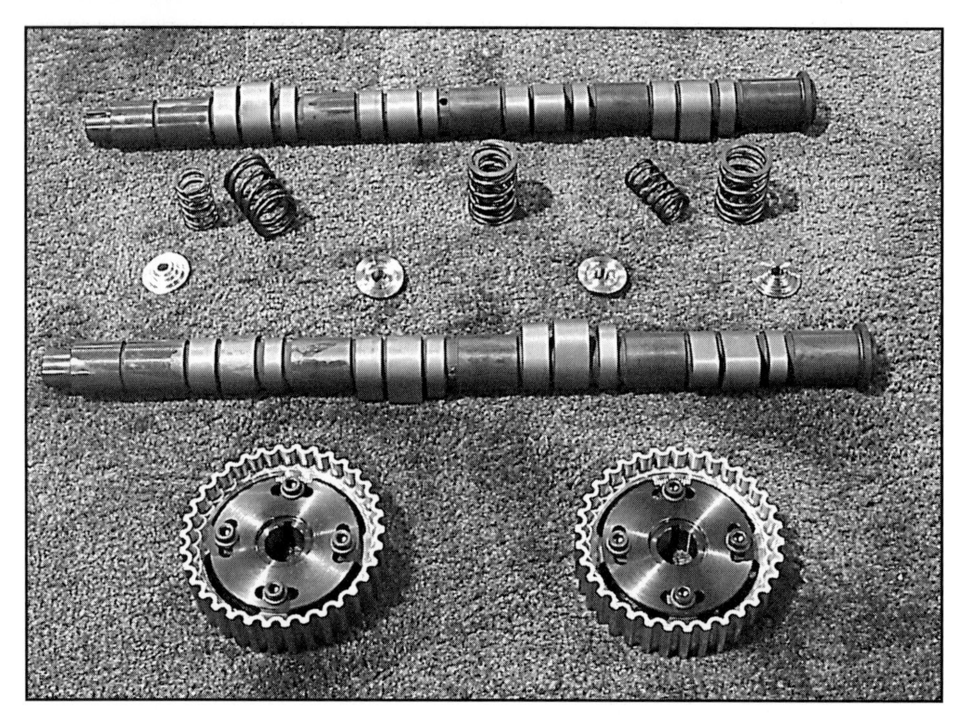

A serious camshaft needs matched components to work properly. A JUN billet B18C cam is shown here with matching adjustable cam gears, valve springs and ultra-light titanium retainers.

half of the crankshaft's speed, driven by the crankshaft, via a belt. The camshafts are in charge of opening and closing your intake and exhaust valves. The cams affect how the air and exhaust gasses flow into and out of the engine. After the basic bolt ons have been installed, the camshaft is the next logical part to upgrade for the reasons of ease of installation and price as well as function. A good pair of cams (or cam in the case of the D15-16) can increase your power more than any other single bolt-on and can synergistically complement your intake, header and exhaust. The wrong cam choice can also turn your car into a poor idling, soggy, slow gross emitter of hydrocarbons.

The camshafts are in a way, the brain of your engine. By controlling when, for how long and by how much the valves open, they control the power delivery characteristics of your engine. If you want a stump pulling torque monster or a high-strung, hyper revving horsepower motor, the cam is the one part of the engine that has the most influence over these characteristics. If you look at a camshaft, you can see that it is a cast metal rod that spins in the cylinder head by its journals, bearing surfaces that are cast into the aluminum head. On this rough cast rod are precisely machined eccentric bumps or lobes that control the opening and closing of the valves. To translate the rotating eccentric motion of the cam's lobes to an up and down motion opening and closing the valves, there is the valve train. For those of you who don't know, the valvetrain are the linkages that transmit the rotary motion of the eccentrically ground cam lobe to the up and down motion opening and closing the valves. These links are typically called rocker arms or cam followers. Imagine a teeter-totter where the cam lobe is on one end and the valve is on the other with a fulcrum in the middle. As the cam spins, the lobe moves the follower up and down which in turn opens and closes the valve. This is more or less how a valvetrain works.

What Does DOHC and SOHC Mean?

Most modern sport compacts are of the Dual OverHead Cam or DOHC design, where there is a separate cam for both the intake and exhaust valves. For high performance use, the DOHC design is preferable because it can use a more compact, simpler valvetrain. The compact drivetrain weighs less, having less mass, a big advantage particularly at high engine rpm. We will explain why in the next paragraph. The Honda/Acura B and H series motors are examples of DOHC engines.

The smaller the moving parts the less reciprocating mass the valvetrain will have. This allows for lighter valve springs, which will have smaller loads on the moving parts, less wear and less friction. Less friction means more power. Less mass also means that the valvetrain can support higher rpms for a given valve spring tension. One of the problems of a valvetrain at high rpm is valve float. This is when the inertia of the valvetrain overcomes the spring tension, which is trying to keep the valves and the cam followers following the contour of the cam. When this happens the engine will misfire and the pistons can even hit the valves causing serious damage. When the valvetrain is bouncing out of control the whole valvetrain is also put under severe stress and wear. Multiple failures of the valvetrain can result. In fact, valve float and valvetrain stability is usually the main limit to the redline or maximum safe rpm limit of an engine. Often when hopping up a motor for the higher rpm that a racing cam can encourage, you must put stiffer valve springs in to prevent valve float.

Hondas use a mechanically adjustable rocker arm end that is threaded with a locking setscrew. This is called a mechanical valvetrain. When the operating clearance of a mechanical valvetrain changes due to wear of the cam, valve, valve seat or follower, a mechanic must adjust the valvetrain's clearances, usually periodically during a tune-up. The mechanical valvetrain is one of the reasons why most Honda motors can attain high rpm operation with very little problems.

SOHC—SOHC or Single OverHead Cam valvetrains have a single, centrally located camshaft connected to the valves with longer rocker arms. The added mass of the longer rockers can reduce the rpm that the valvetrain can reach before float. To counter the mass a stiffer valvespring with more resulting friction and wear can be used. The Honda Civic D15 and D16 are good examples of the SOHC designs. The Accord F-series motors are also SOHC. A SOHC is still better than OHV or overhead valve, valvetrains like those found in domestic V-8's. These archaic pieces have a single cam deep within the engine block with long pushrods to connect the cam's lobes to the rocker arms. These valvetrains have a lot of flex and inaccuracy in them in addition to a lot of weight. This limits them to low rpm with heavy wear because the valve springs are very stiff, which creates a lot of friction.

So as a short recap, DOHC valvetrains are the best for high rpm use, SOHC are next best, and

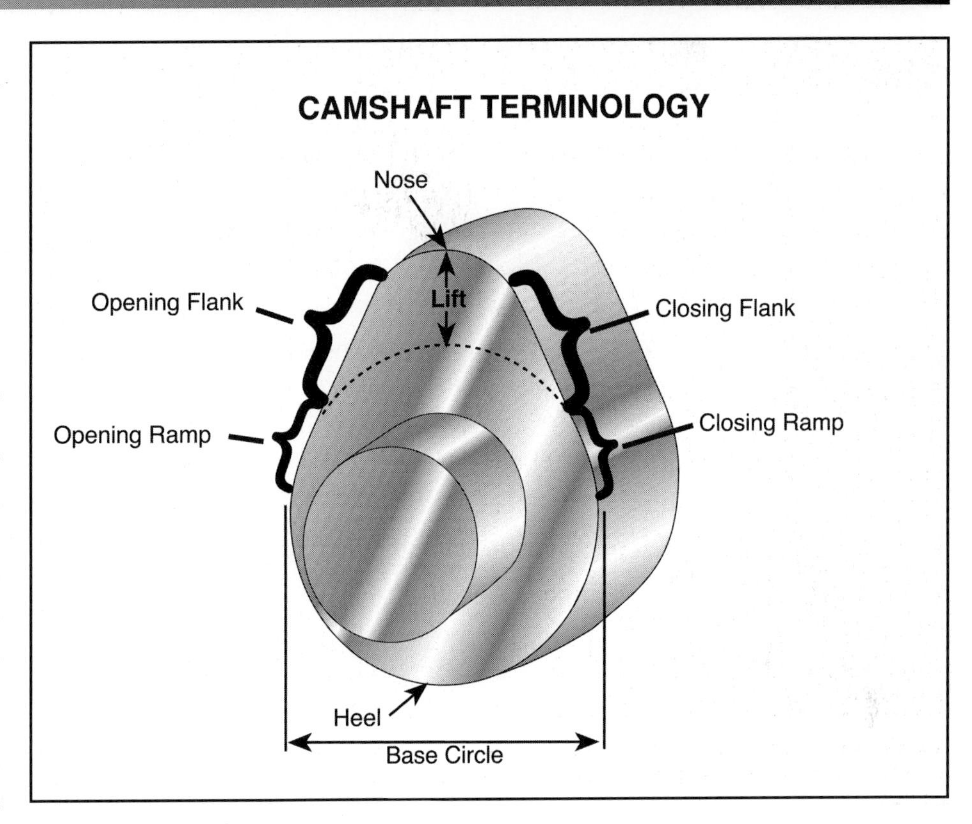

Overhead Valve are the old-school throwback.

Cams Specs or "What do Those Weird Numbers Mean?"

Now that you understand the monkey motion of the valvetrain, let's talk in-depth about the cam or cams in the case a DOHC motor. There are four main things to look at and understand when buying a cam: Duration, lift, overlap and the closely related lobe separation angle. Duration is the number of degrees of crank rotation that the valve is held open by the cam. Remember that the cam spins at one-half engine speed and that the total 4-stroke cycle is 720 degrees or two complete rotations of the crank. As a general rule of thumb, the more duration a camshaft has, the more top-end power the cam will help the engine produce at the expense of bottom-end power. With less bottom-end power, part throttle driveabilty becomes worse with flat spots in the power occurring in really big duration race cams. The idle speed becomes higher and the idle itself becomes rougher with a rolling lope to it. We will explain why a little later. The more duration a cam has, the more degrees of crank rotation that the valve is held open. The longer the valve stays open, the longer time is available for the cylinder to be filled. This is important at high rpm as there is less and less time to fill the cylinder, the faster the engine spins.

Lift

Another factor is *lift*. Lift is the height that the valve is lifted off of the valve seat. Usually the more lift within reason the better as the higher the valve is lifted, the more flow that can go past it. It's like holding your door open on a windy day, the wider you open the door, the more wind blows into your house. The only drawback to having a lot of lift is that the valve opening and closing speeds become high, increasing the risk of valve float. Valve float is when the opening and closing speed of the valves becomes so fast that the springs cannot keep up with the valves motion, causing the valves to float away from the cam's profile. This can cause serious damage if the floating valves smack the pistons. The valvetrain's natural frequency can also come into play here as sixth and seventh order harmonics can also contribute to the forces that cause valve float.

Harmonics result in the build-up of forces caused by the cyclical motion of the same frequency. When many small forces are added together as a result of the harmonics, they can become significant and damaging.

To combat valve float, stiffer valve springs must be used. Springs with a very high natural frequency are also used in well-engineered systems. Stiffer valve springs create more friction within the engine and promote better wear of valvetrain parts. Lightweight titanium valve spring retainers reduce the inertial forces that contribute to valve float. Because of these issues, there is a limit to how high a valve can be lifted before the point of diminishing returns. If only a valve's lift is changed with no change to the duration, usually more top-end power is generated with only a little loss in bottom end. In effect the powerband becomes wider and more useful. Unless great care is taken in the design of the valvetrain and springs, the wear can become extreme and the service life can be very short due to the high velocities and thrashing that a high-lift, short duration cam can cause. The lighter the valvetrain the better in this case. Usually the amount of valve lift and duration is compromised for when better durability and low-end power are needed, like in a street-driven car.

So for more top-end power, you want a cam with long duration and as much lift as possible. However, like we said before, a cam like this is not the best for most use other than full-on hardcore racing. Here is why: As the duration of the intake and exhaust cams is increased, the points of the cam's opening and closing are pushed out. The valves open earlier and close later. When the intake valve opens earlier and earlier, it gets to the point where the intake valve is opening when the piston is pretty far down the bore coming up to but not quite on the intake stroke. So the intake valve is opening while the piston is still coming up. At high engine speeds when the gas column of rapidly moving air in the intake port has a lot of inertia because of the almost constant but cyclical gulp of the engine sucking in a lot of air, this is a good thing giving more time to fill the cylinder. Unfortunately all good things must come to an end. When the engine speed slows down, the incoming air looses inertia and the rising piston can push air out of the cylinder back up the intake port. This is the opposite of what we want to happen as it seriously hampers the ability of the engine to flow air at lower speeds. This phenomenon is called *reversion* and is one of the main reasons why racing cams lack low-end power.

Blowdown

When the exhaust cam has its duration increased to longer and longer openings to help give the engine time to expel more of its exhaust at higher rpms, the opening of the exhaust valve is moved earlier and earlier. Pretty soon the exhaust valve opens when there is still a bunch of pressure left in the cylinder from the end of the power stroke. This pressure is immediately bled off and shot down the exhaust pipe. This is called *blowdown*. Blowdown causes the engine to loose some of the power from the power stroke. When the engine is running at high rpm and needs all the help it can get to fill and exhaust the engine as quickly as possible, some blowdown is permissible to help breathing. However at lower engine speeds, excessive blowdown causes a loss of power by prematurely venting off work-producing combustion pressure.

Overlap

The other main factor in cam design is *overlap*. Overlap is the period at the end of the exhaust stroke and the beginning of the intake stroke when both the intake and exhaust valves are open. Overlap is important because having both the intake and exhaust valves open at the same time causes better scavenging of stale exhaust to occur. This is because the gas column inertia of the inrushing intake charge will cause it to flow across the cylinder to do a more thorough job of pushing the stale burnt exhaust gas out of the cylinder. Obviously, too much overlap will cause the intake charge to rush out the exhaust port, wasting fuel. Another drawback to overlap is that it can also contribute to reversion in the intake port as explained a few paragraphs earlier. The reversion causes intake charge dilution at low rpm as the backflow in the intake ports hampers cylinder filling. This does two things, the cylinder pressure is poor at low speeds because of incomplete filling

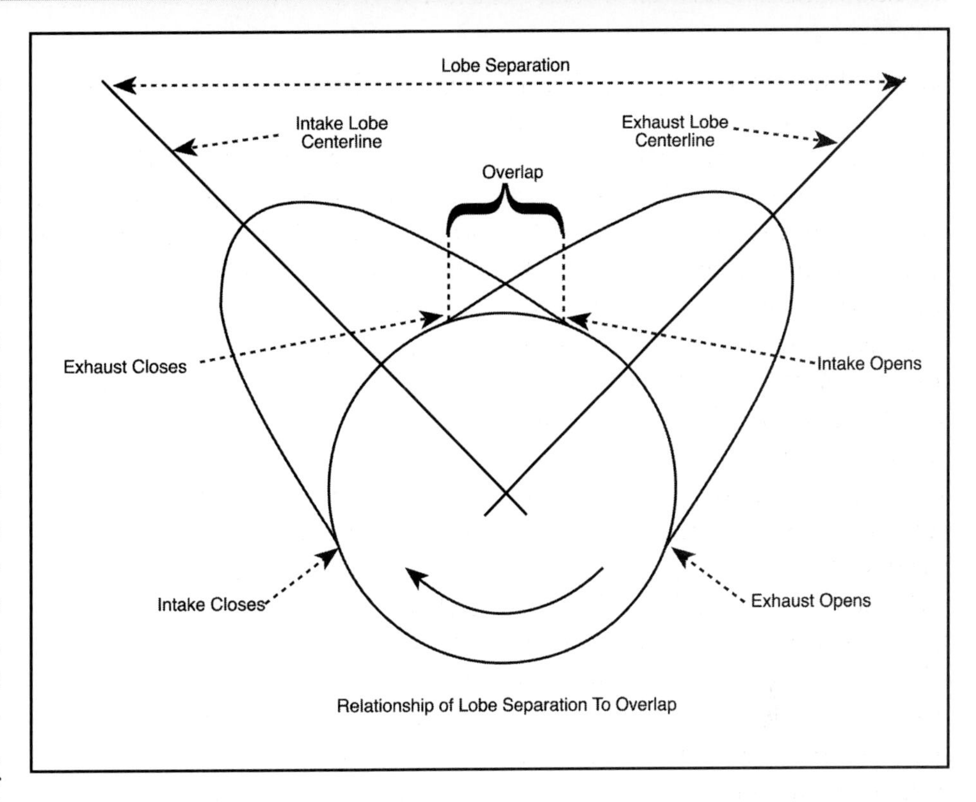

Relationship of Lobe Separation To Overlap

and the fuel/air ratio of intake charge gets diluted because of the air rushing backward up the intake port causing a low speed misfire. These poor conditions for combustion causes the engine to skip a beat in a rhythmic fashion at idle and low speeds usually firing once every four revolutions when the cylinder gets enough fuel to touch off. This phenomenon is called *8-stroking,* because it is pretty predictable and rhythmic as the engine steadily skips beats at low rpm because it is firing every other stroke of the 4-stroke cycle. This is why racing cams have a distinct lopey lub-dub idle. Although performance nuts find this a neat feel and sound to have, non-performance folks hate the rough idle, and the misfire causes lots of hydrocarbon pollution. Because of today's strict pollution requirements, most of today's stock cams have very little overlap.

Lobe Separation Angle

Lobe separation angle is the number of degrees that the points of highest lift on the intake and exhaust cams are separated by. This is a useful number for two reasons. One, when you are dialing in your cam timing, which is an operation that should be done when you build a high performance motor, you need to check to make sure that your cam is ground correctly. Due to the gradual nature of the cam's lobe when transitioning from the round basecircle to the valve opening slop of the lobe, it is hard to find the exact valve opening point of a cam even with precision dial gauges. If you know the cams Lobe Separation Angle, which should be spec'ed by the cam's manufacturer, you can degree in your cams. This is done using a degree wheel indexed at zero with a pointer at TDC on the engine's crankshaft and having a dial indicator on the valve. The engine is turned over in its direction of normal rotation (clockwise in most engines, counterclockwise in most Hondas) until the valve is at about 90% of its rated lift. Note the indicated degree of the crankshaft at about 90% of its lift, then continue to turn the engine over until the cam spins over the top of the lobe until the 90% lift point is met on the other side of the lobe. The degrees of the two points should be added and divided by two. This will give you the lobe top center in crankshaft degrees of one cam. The process should be repeated on the other cam. The number of crank degrees that separate lobe top centers of the two cams is the lobe separation angle. For motors with hydraulic lash adjusters, the adjuster must be disabled somehow as the mush in the lifter will cause inconsistent readings. However, popular Honda motors have mechanical lash adjustment. For obvious reasons, it is a lot easier to do this when the motor is first being assembled, out of the car.

It is important to degree-in a cam because many built engines have had the block decked or the head milled during the building up process. When material is removed from the head or block, the distance between the crank and cams is reduced, which puts more slack in the timing chain, retarding the camshafts, which could result in a loss of power. With adjustable cam gears, the cams can be corrected for proper timing. Unfortunately it is very common for aftermarket performance cams to be ground slightly off index. That is because many cam grinders do not use import specific tooling to index the cam blanks to their grinding machines using adapters and other jury-rigged methods. Most cam grinders have the stuff to do small-block Chevy stuff correctly but since import stuff is new and lower volume, the tooling for them is often improvised. This results in the cams being ground up to several degrees off center. Even famous name cams are sometimes ground off. That is why it is important to double-check your cam when it is installed. That is why cams tested in some magazines often discover huge power gains when adjustable timing gears are installed in their project cars. They are actually correcting for mis-indexed cams in the first place! That is why many tuners often say that every engine requires different settings of the adjustable timing gears. It is not the engine so much as the camshafts. Not too many shops or even magazine editors understand how to degree in camshafts or why it is important but checking cam timing is standard operating procedure in real professional racing shops.

We will discuss the quick and dirty way to dial in cams on a dyno once the engine is in the car or if you do not have the precision instruments like a dial indicator toward the end of this section but this background information will help prep you for it.

The second reason that Lobe Separation Angle is important is that you can understand some of the cam's behavior by looking at it. Lobe separation angle has a direct correlation to the amount of overlap an engine has. A smaller Lobe Separation Angle has more overlap. A bigger angle has less. The amount of overlap has an effect on the engine's powerband just like lift and duration do. Old-school V-8 motors with poor breathing 2-valve OHV engines need a lot of overlap to get the flow going at high rpm. It was kind of a crutch to get around the poor layout of the ports and valves. An old-school V-8 needs lots of duration and a tight lobe separation angle. A hot street Small Block V-8 cam might have something like 270–290 degrees duration with a lobe separation angle of 106 to 112 degrees. That is why many V-8's lope like crazy and sound so mean when they cruise the burger stands. In fact, lobe separation angle is one way to figure how badly an engine might idle. The smaller the number, the more overlap and the more 8-stroking and lope. Buy a tight lobe separation angle cam or in the case of a DOHC, adjust them tight and intimidate the boys at the drive-in!

Now our modern port layout, free-breathing, high-revving 4-valve-per-cylinder, low-included valve angle Honda motor typically has a much better breathing cylinder head than an old-school V-8. I mean the current popular Honda engines were designed relatively recently and our V-8 brethren were conceived way before even an old guy like me was born. Thus, our motors do not tolerate or need as much duration and overlap as the old-school V-8 to breathe at high rpm. A hot street cam for a Honda has something like 260–280 degrees of duration with a lobe separation angle of about 112–118 degrees. We simply do not need as much overlap and duration to make the high-end power. This is considering that a cam with these sort of specs can easily pull up to 8000 rpm on a well-built sport compact motor where the equivalent cam on a V-8 will wheeze up to 6500 rpm tops.

So if you understand all of this, you know that a high lift long duration set of cams have better power at high engine speeds but a poor idle and bad emissions at low engine speeds. For most of us this means that a good choice for street driving is somewhere in the middle of this, getting a good compromise between the smooth idle, clean burning and good low end torque of a stock cam and the screaming high rpm breathing of a race cam.

The exception to the rule is motors with variable cam timing. These motors can change cam spec while the engine is running to try to keep things optimum across a wide range of rpm with little compromise. The famous Honda VTEC system is the premier example of variable cam technology in the automotive world to date.

What is VTEC?

The best way of exploiting a changing cam spec is the Honda/Acura VTEC system as found on the B16A and the B18C. This revolutionary, amazing system uses two different cam lobes. At low speed, a short duration, low lift, wide lobe separation angle cam lobe is used. At a predetermined point, the engine's computer activates the high rpm cam which is at near race car specs. The high rpm lobe on an Acura Integra B18C5 motor has a nearly full-race duration of about 290 degrees! (or about 242 degrees @ 1mm) This is the best of both worlds, a smooth idle, good emissions and bottom-end power with a killer top-end charge. VTEC is Honda's killer advantage and the main reason why Honda motors have the highest Hp to Liter of any production naturally aspirated automotive engines.

The VTEC system on the D-series motors found on Civics is not as sophisticated, it merely advances and retards the cam to widen the powerband, and it does not have all of the advantages of having two different sets of cam lobes as the B16A and B18C does.

The latest Honda/Acura motor, found in the Acura RSX Type S and the Civic Si, use the latest variant of VTEC or VTECi. VTECi combines the features of both the B-series and D-series VTEC for even greater powerband width and cleaner emissions.

Why Do Duration Figures Vary So Much?

This is because there is no standardized way of measuring duration in the industry. The figure you most often see advertised or what the salesman might tell you is called Advertised Duration. Usually but not always, this is the same as SAE (Society of Automotive Engineers) duration. This is duration measured on the cam lobe by dial indicator and degree wheel when the dial indicator reads 0.003" lift. Why 0.003"? It is because it is really hard to see when a dial indicator first starts to move vs. degrees on a degree wheel when measuring a cam. The opening point is spread out over quite a few degrees and it is pretty hard to nail the exact opening point. 0.003" is an easy spot to pick where you can get a definite reading.

Sometimes cam makers will state duration differently. Commonly, the makers of domestic cams will pick 0.050" as the measuring point for the cam's duration. This is because some engine builders feel that the sloppy pushrod and floating rocker arms that domestic engines sport add a lot of slop into the equation and 0.050" is a more accurate point to take a cam's duration from. Some Japanese import cam makers use 1mm or 0.040" for the same reasons. Honda uses 1mm for their measuring clearance. Duration taken from these points of higher lift will be quite a bit less than advertised duration. A cam with 260 degrees of advertised duration will have 0.050" duration of around 215 degrees. Some cam grinders use 0.030" or even something like 0.020" because they feel that that is the best place to measure duration for no reason but their own. So when buying a cam, find out how the duration is measured. For instance, a cam with 260 degrees of advertised duration is a good mild street cam. A cam with 260 degrees of duration @ 0.050" lift is a full race monster!

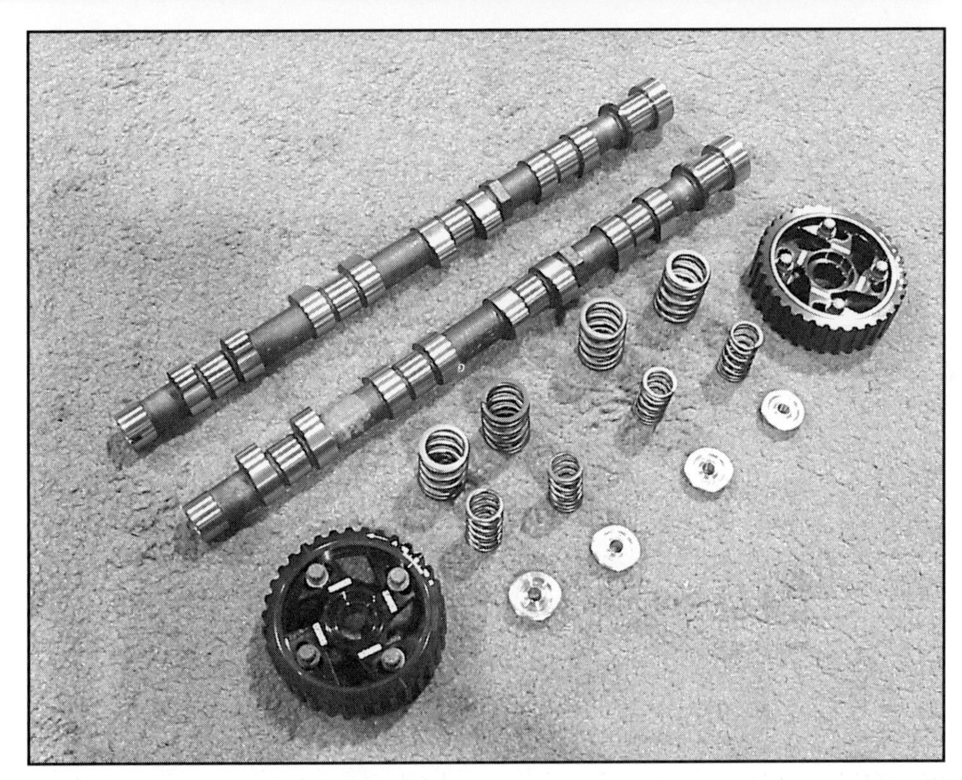

The Crower billet cam set gives the common Integra base motor, the B18B the punch of its famous brother the GSR B18C. Here is the Crower cam with matching valvesprings, lightweight titanium retainers and adjustable timing gears.

What Cam Will Work for Me?

Here are some very general typical specs for different types of cams. These are very general guidelines for a modern 4-valve per cylinder DOHC 4-cylinder engine of between 1600 and 2000cc's. The durations stated here are Advertised Duration.

240 degrees of duration, 0.390" lift, 15 degrees of overlap—A stock cam, works well from 700–6500 rpm, smooth as can be. Good pollution.

265 degrees of duration, 0.420" lift, 30 degrees of overlap—A good street cam, pulls good from 4000–7500 rpm, slight lope to idle. Idle at about 900 rpm. Might pass a smog test. Might work with OBDII. Works well with most bolt-on power mods like header, exhaust, air intake. Car feels fun to drive on the street and is most likely more responsive than stock almost everywhere in the powerband. Most people should stop here.

280 degrees of duration, 0.440" lift—A mild race cam, pulls good from 4500–8000. Idle at 950 rpm with a definite lope. Could pass a smog test with tricky fuel and ignition management, most likely OBDII will not like this. Car might not feel too good from a seat-of-the-pants perspective. Higher compression (10:1), headwork and headers become a big plus at this point. Special matched valve springs are most likely needed.

290 degrees of duration, 0.460" lift—A high rpm lobe of a VTEC cam (duration only) really wild street cam or mild race cam, powerband from 5500–8500. Idle lopy at about 1200 rpm. Most likely will fail HC and CO part of smog test. OBDII system may experience some serious emotional problems. Higher compression (11:1), headwork, headers and other stuff become almost mandatory here. Racing valve springs are needed. Compression puts motor almost to the point of not working with pump gas.

305 degrees of duration, 0.500" lift—A full on race cam, powerband from about 7000–9500 rpm. Idle lopy at about 1400 rpm. Smog?? Hah! OBD II system giggles and spews failure codes before becoming catatonic. High compression (at least 12:1), headwork, short runner manifold, custom tuned header, bigger injectors, high-powered ignition and stiff valve springs become mandatory. Car won't run on pump gas anymore because of compression. Close ratio transmission needed due to narrow powerband. Car is just about undrivable on the street unless it's certain streets in Long Beach in April. Some cams are even bigger than this. This author has used cams with durations of up to 320 degrees on some race cars.

Camshafts and Turbocharged Motors

Turbochargers add backpressure to a motor because having the turbine in the exhaust stream that is recovering energy to spin the compressor will cause some exhaust gas to back up in the exhaust manifold. This backpressure can range from a negative pressure where there is more boost than backpressure, which is possible for racing application big turbos to situations where there is as much as three times more backpressure than boost pressure, in the case of small turboed street set-ups. The amount of backpressure that a turbo creates will have a big influence on what kind of cam profiles and cam timing the motor will like.

Generally, a streetable mid-size turbo like the Garrett T3 turbine T04 compressor hybrid, which is popular with many Honda turbo kits, will run a backpressure to boost pressure ratio of about 1.5:1 at around 15 or so psi. This can vary somewhat due to difference in engine mods, Turbo A/R and trim, boost pressure run and RPM but is a fairly good general guideline. A turbo like this with mild amounts of backpressure will generally prefer near a stock cam, especially for the V-Tec equipped motors with their bigger secondary lobes. The D-series SOHC motors will prefer cams of under 270 degrees duration with a lobe separation angle of 114 or more degrees.

The reason why turbo motors do not like lots of duration and overlap is because of the backpressure issue. If there is much more backpressure than boost pressure, it is possible for hot exhaust gas to flow backward through the motor, contaminating the intake charge and allowing heat to build up in the motor. Normally at higher rpm where the bad effects of overlap (poor idle) are eliminated, the backpressure is greater as the turbo works harder making the problem worse.

Smaller turbos that some kits offer like the Garrett T25 series or the Mitsubishi TDO5 that are sized for less lag, have even greater issues with backpressure, especially when the boost has been turned up past the kit's base settings. It is possible to see as much as 40–50 psi of exhaust manifold pressure under some conditions with these smaller turbos. With high manifold backpressure, it is important to run a stock cam with minimal overlap.

With adjustable timing gears it might be beneficial to retard the intake cam and advance the exhaust cam to reduce overlap. This is a worthy experiment with any turbo car. It usually hurts off boost response less if you experiment with advancing the intake cam a few degrees first before retarding the exhaust cam. A side benefit of reducing overlap is going to be a smoother idle. Increasing blowdown this way may also spool the turbo faster.

Some of the big, drag-only turbos like the Turbonetics T-series turbo, when paired to big turbines, can get away with more radical cam timing due to lower backpressure. If you are building a maximum effort turbo motor that is not intended for street use and has an exhaust manifold pressure close to or less than the boost pressure, you can run a cam spec'ed pretty close to what a hot naturally aspirated motor would run for your intended RPM range. If you have adjustable timing gears, which are definitely recommended for all serious motors, you may want to experiment with reducing overlap.

In summation, most hot street and larger turbochargers like mild camshafts with only a few degrees more duration than stock with the lobe centers adjusted to have a near stock lobe separation angle. Engines with turbo kits with small turbos are usually better off retaining the stock camshafts, and V-TEC DOHC motors also usually do better with stock camshafts.

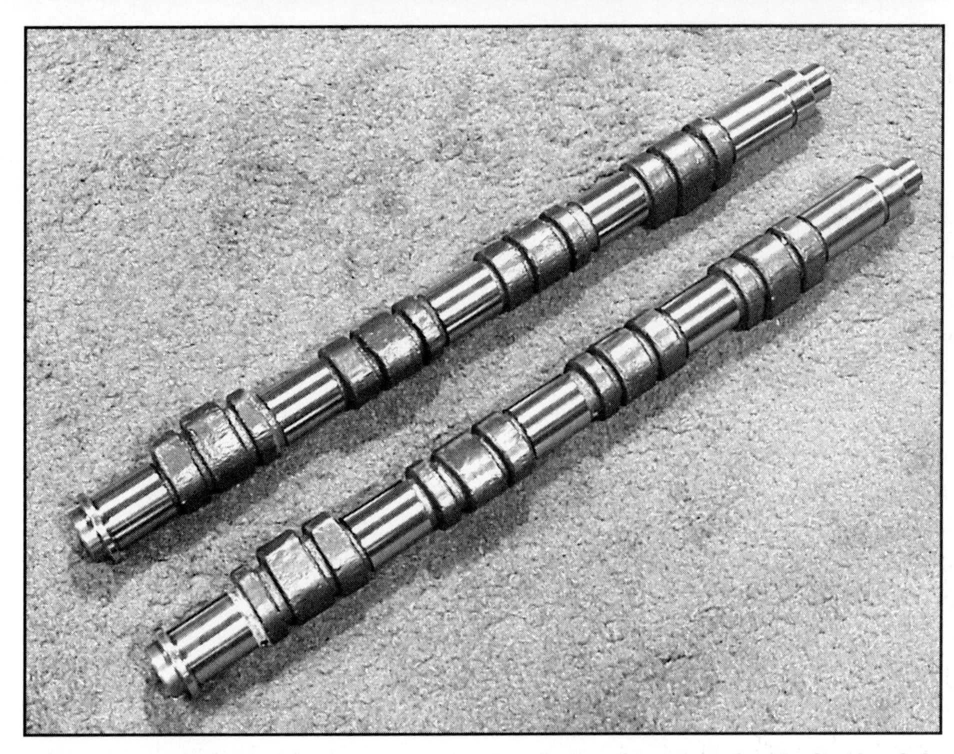

Although they can work well in some motors, like direct-acting valvetrains, for Hondas, it is best not to consider regrinds when choosing a cam. Here is an unground blank billet. Notice how big the unfinished lobes are. A billet cam maintains proper rocker arm geometry for best power and long-term wear.

Camshafts and Supercharged Motors

Lots of duration and overlap are to be avoided in supercharged applications also. If too much overlap is run in a supercharged engine's cam profile, some of the boost pressure may be blown out the exhaust manifold! It is not unusual to see a slight reduction of boost pressure when installing a cam with a supercharged motor. Driving the supercharger slightly faster can compensate this for. Like the turbocharger, the supercharger also prefers near stock mild or stock camshafts but for a different reason.

Camshaft Choices

When buying a camshaft, it is best to buy billet camshafts. A billet camshaft is ground on a new billet with a base circle diameter the same as a new stock cam. Unlike your typical small-block Chevy or 5-liter Ford, the cam aftermarket has not caught the import performance bug to the same degree (the Ford and Chevy guys have been at it 30 years longer!), so cam choices are few and far between. Many cams available for Hondas, especially the Civic D-series motors are low production and done as regrinds, because the volume does not justify the tooling to make billets. As we have said before, there is no guarantee that this grinding is indexed properly as this is difficult for a cam grinder to do with makeshift fixturing. When a cam is reground, metal is removed from the cam's base circle or bottom, which changes the relationship of the base of the cam to the top of the lobe. This gives the cam grinder an effectively bigger lobe to regrind. A regrind camshaft is made to have more lift and duration by grinding on the base circle of the camshaft in this way. Although this trick works to an extent, motors with finger follower valvetrains like Hondas have trouble with maintaining valvetrain geometry with regrinds. A regrind will not measure out like how it is spec'ed unless a special asymmetrical lobe contour is used. Not too many small cam grinders that typically do regrinds have the technical resources to properly design a profile like this. The reason why a regrind will not measure truly is that since the base circle of the cam is now smaller, the contact patch of the cam follower is now contacting the lobe in a different place for where it was designed to contact, which throws off the rocker ratio and makes the valve move slightly out of sync with the lobe of the cam. It is not uncommon for many poor quality aftermarket Honda cams to be ground off by a few degrees. Because of these reasons, cars equipped with regrind cams can benefit greatly from a dyno tuning session with some adjustable cam timing gears. Because of these issues, we don't recommend regrind cams ever on Honda motors, not even from makers with supposedly good reputations.

For the VTEC B-series motors, the stock cam is pretty hard to beat. The

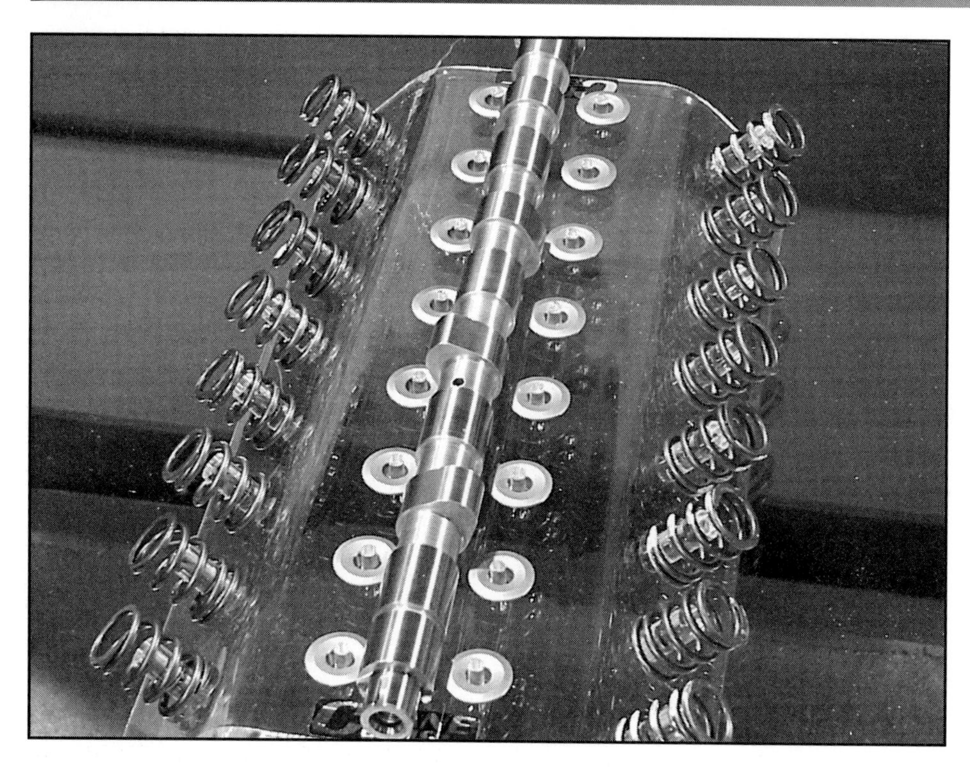

Crane's roller cam for the B18C and B16A forgoes the low rpm VTEC lobe, making it more of a full-race only setup.

high rpm lobe of these motors is just about as radical as many full-race cams. The stock Type R cam can be retrofitted onto the B18C and the B16A motor and is a very good choice; on a street motor, it is a better choice than many aftermarket cams. The intake cam from a JDM Civic Type R is the biggest of the factory cams and is also a good cam to use for a mild B18C or B16A buildup. Be sure to switch the intake valve springs to the exhaust side and to use Integra Type R intake valve springs when using the Type R cams. Type R cams are good for 6–8 hp over stock. The H22 VTEC Prelude motor can also benefit for the JDM Type S cam. They are good for 3–5 hp. The aftermarket cams for these motors are often beneficial only if the motor is going to be modified enough to rev in the 9000-plus rpm range.

Toda, Skunk 2 and JUN make excellent billet camshafts for the V-Tec B16A and B18C motors while Crower and Crane has good grinds for the Non V-Tec B-series DOHC motors. The D-series SOHC motors have cams done by Crane, Crower and JUN.

High Tech Exhaust and Crane are currently offering roller cams for the B-series engine. Roller cams have a roller on the end of the cam's follower instead of a rubbing block. The roller reduces friction and more importantly allows a much more radical lobe profile for more area under the lift curve. This gives more top-end and mid-range power with the same duration, thus not affecting the engine's bottom-end power as much. The new roller cams look especially

High Tech Exhaust's B18C and B16A roller cam is a very neat unit, having a radical roller profile only on the high rpm VTEC lobe. High Tech's system still has the advantage of functioning low rpm optimized lobes.

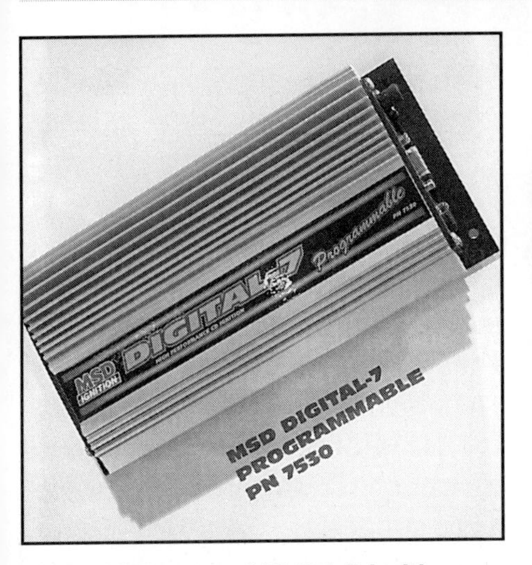

MSD's high-powered Digital 7 ignition can ignite cylinders pressurized by lots of boost pressure and NOS. The Digital 7 has built-in features that a racer could appreciate like a two-step rev limiter and a soft-touch rev limiter as standard equipment.

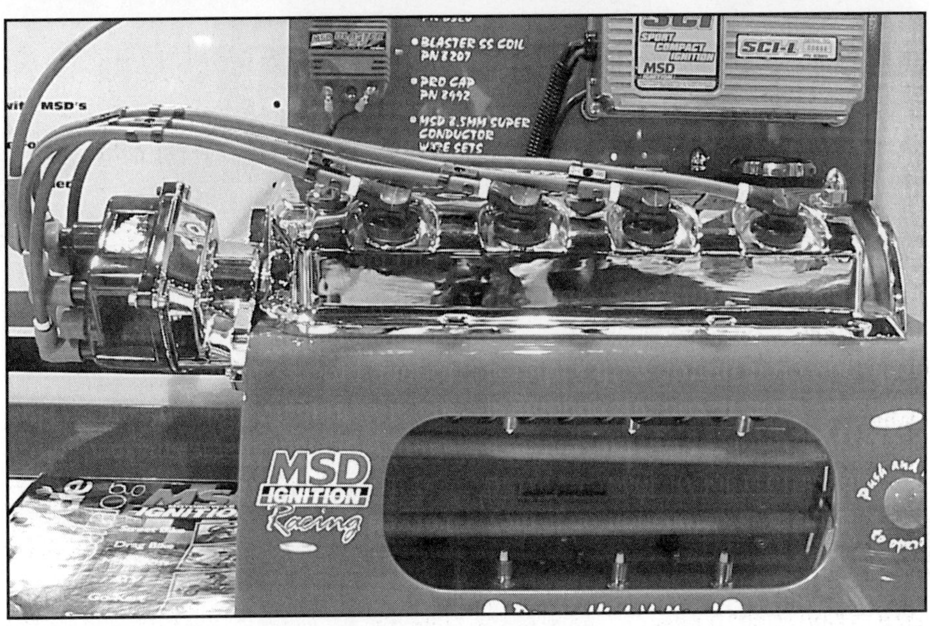

MSD's SCI ignition is a powerful, affordable unit for street use.

promising.

A very important note is that when changing camshafts be sure to use without fail, the valve springs recommended by the cam's manufacturer or severe problems will surely result. Valve float and coil bind can result by using the wrong springs. This will destroy the engine's entire valvetrain! When installing a camshaft, all of your previous optimized tuning specs as far as cam timing adjustment, ignition timing and fuel pressure may change, sometimes drastically. It is now time to invest in another dyno tuning session. Again the rewards will usually be well worth the cost.

When it comes to camshafts, there are so many different engine combinations that can affect exactly how your engine responds to changes in camshafts and cam timing; there are no hard and fast rules, except perhaps, don't overdo it. A common beginner mistake is to install too big of a cam. Street-driven engines should stick to the factory Type R variety of cams or some of the aftermarkets milder offerings for best results. The stuff above is a guideline on where you can start and what directions you can try. We would also like to stress the importance of actually testing and tuning your cam combination for best results. We will cover specific camshafts in our hot street engine section.

High Powered Ignition Systems

As we are talking about basic bolt-on mods at this point with no internal work at this point, the stock Honda ignition will be fine. The stock Honda ignition is very powerful and modifying it at this point is simply a waste of money.

When turbocharged or supercharged boost, nitrous, or high compression is applied, the benefits of the stronger spark of a high-powered ignition, such as the MSD become pretty apparent. At this point, the best thing you can do is to invest in some high-quality spark plug wires and make sure your ignition cap and rotor are in good condition. Surprisingly the stock plug wires work exceedingly well but if you must have colorful aftermarket wires, make sure that you get spiral wound suppressor core wires. Solid core wires create electromagnetic interference that can wreck havoc with your car's electronic control systems.

This concludes our section on the basic bolt-ons. If done correctly with matching parts, all tuned to work correctly together, gains from 20–50 hp over stock are possible, all without going inside the motor. To get more power than this, it will be time to delve deeply into the engine's internals. In the next chapter we will investigate the secrets of headwork and what it takes to build a bulletproof bottom-end for your super motor.

3 CYLINDER HEADS

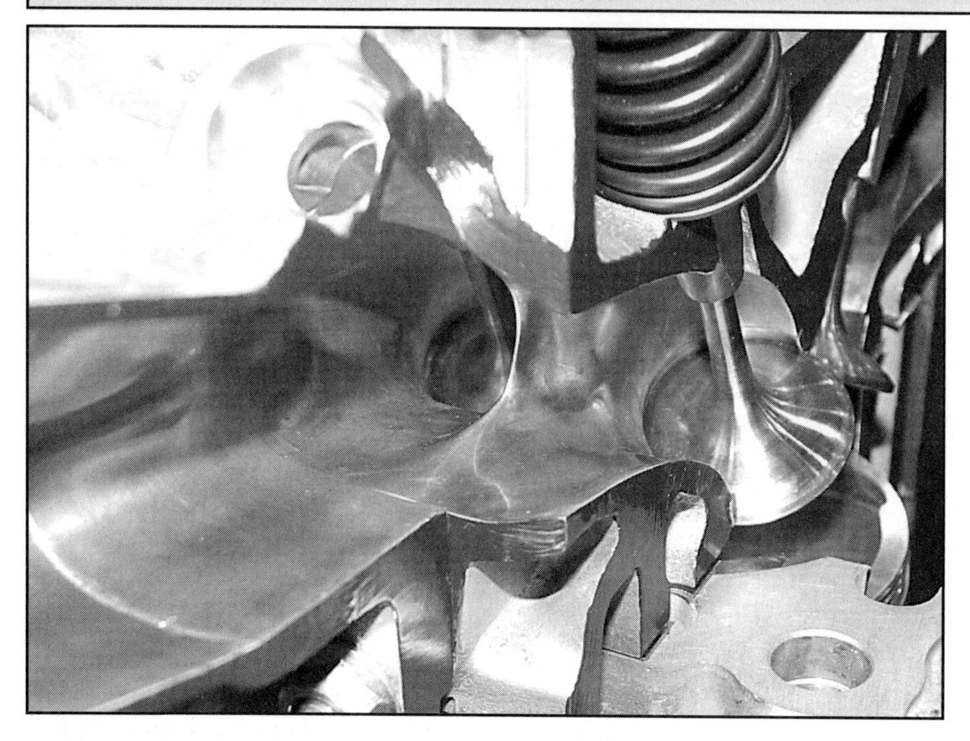

This cutaway of a JUN head shows the intricate hand work that goes into a professionally prepped cylinder head. The ports are enlarged with more material being removed from the roof of the port. The port splitter is moved back and carefully reshaped, the valve seats match the port perfectly, the lightweight titanium valves are contoured, a blended multi-angle valve job is used and a nice smooth finish is applied to the port walls. The fact that this fine piece was ruined to make this demo head almost makes you want to cry!

CYLINDER HEADS

When it comes to building up an engine for more performance, you are already one step ahead of the game with a Honda engine. Honda engines, especially the hot performance motors like the B18C, B16A and the H22, come stock with some of the best ports of any production car, period. So if you are building up one of these motors, you already have a good foundation.

Even the SOHC D series motors as found on the Civic and the lesser B series motors like the B18B non-VTEC Integra motors have fairly good heads to start with. In this chapter, we will cover the ins and outs of cylinder head modifications, arming you with information on headwork techniques and terminology to help you make intelligent choices.

AIRFLOW BASICS

Achieving good airflow through the cylinder head is essential for creating big horsepower. You are probably wondering how increasing airflow gets you more power. To start with, an engine is basically an air pump. To create the most power possible, it is important to get the maximum amount of fresh air into and burnt exhaust out of the engine with the least amount of mechanical effort. Excess intake and exhaust port restriction makes more work for the engine, work that is not used to turn the wheels. This extra work is called pumping loss. Imagine sucking a thick milkshake through a small straw. This takes a lot of sucking effort on your part. Now drink that same milkshake through a larger straw—it's easier isn't it? The extra effort you are expending to drink that milkshake through the smaller straw is a pumping loss. The more work it takes to get the various gasses into and out of your engine, the less power will be available to spin the crank and the wheels.

Volumetric Efficiency

Another major factor is volumetric efficiency. Volumetric efficiency is the percentage of an engine's displacement that is filled on each intake stroke. To make things easy to understand, let's imagine that you have an engine that displaces 1000cc. If the engine can take in 800cc of air on the intake stroke, the engine has 80% volumetric efficiency. Eighty-percent volumetric efficiency is typical for a modern, production-based naturally aspirated engine.

The factory, especially Honda, carefully calculates an engine's port size and configuration to optimize volume and geometry in stock configuration. However, when an engine is modified for more power,

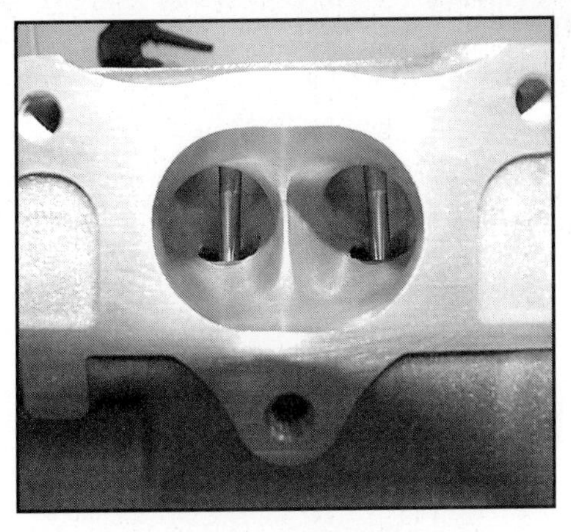

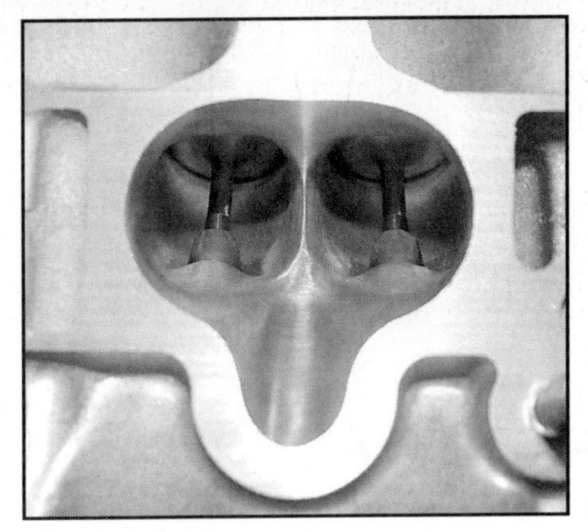

This fully modified B18C cylinder head by Port Flow Design shows all the tricks of the trade; polished combustion chambers, polished and reshaped intake and exhaust ports, big titanium valves and a multi-angle valve job. This head was used on an engine that set the IDRC all-motor record with an ET of 10.79! Believe it or not, this head would also work well on a hot street motor!

the flow demands of the engine also change. Air intakes, headers, exhaust systems and cams that promote higher rpm operation also demand more airflow. More flow often requires bigger ports with more cross-sectional area. Remember drinking that thick milkshake through a straw twice the diameter? That is the main effect of porting; reducing pumping losses and increasing volumetric efficiency by reducing restriction.

Sometimes the design of the head and the ports layout itself is compromised to make the head fit for a lower hood height or even bending around head bolts, water passages or pushrods. It is only recently that engine designers, specifically the designers of high output import engines, designed the head first around the ports for best flow and good volumetric efficiency. Domestic motors seem to place the importance of the location for almost every component before the ports. The designers of the Big Three American manufacturers seem to place the ports wherever there is room in the casting, considering their position last in the design process, long after all the other parts of the head have been located. This is partly because many of these motors were designed a long time ago and the manufacturers are trying to maximize profits from the tooling built for these old designs.

PORTING

In cylinder head porting, the intake and the exhaust ports of the head are carefully reshaped by hand, enlarging, straightening and streamlining them to reduce any pumping-loss inducing restriction and to reduce turbulence to increase flow velocity as much as possible. For the most part, the ports are straightened out using a die grinder and a carbide bit to a line of sight configuration. Straightening gets rid of turbulence-producing bends in the port. Porting involves extensive hand finishing with the aforementioned die grinders to remove tooling cuts, sand casting pits, and the usual lumps and bumps made by the mass-production tooling of the

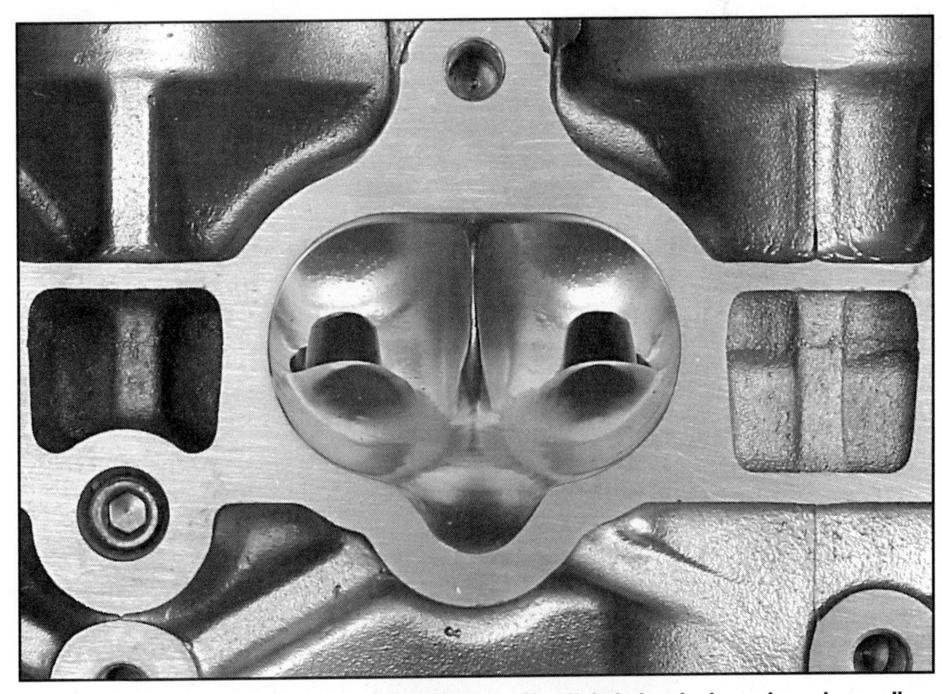

Extrude Honing is very effective on cylinder heads. Lisa Kubo's head, shown here, has wall so smooth that they appear to be polished. Her motor makes over 650 hp at the wheels!

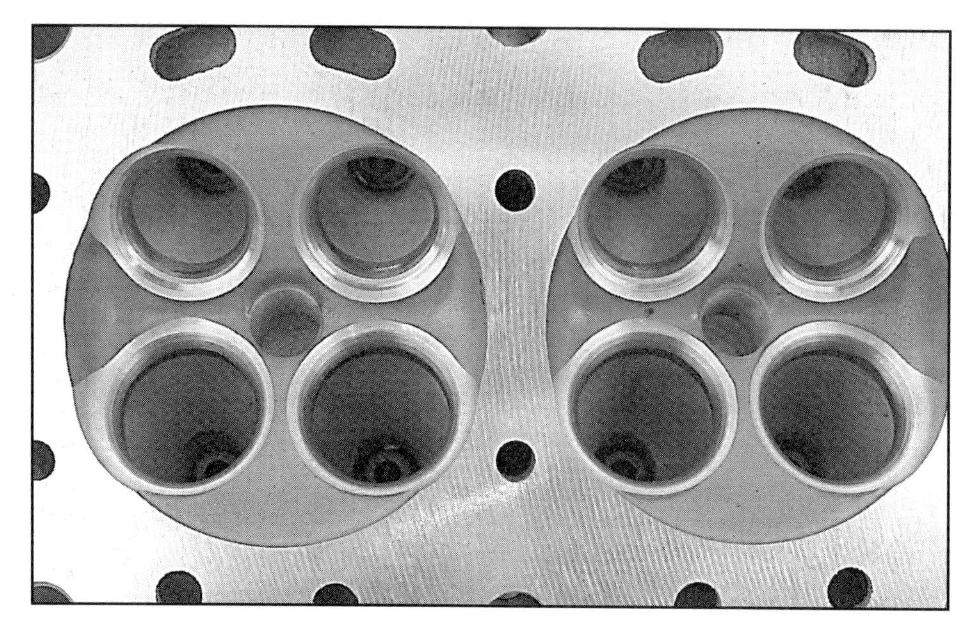

This D16 cylinder head by RS Motorworks shows some unshrouding of the valves, the shiny machined areas between the valves and the edge of the combustion chamber, a good multi-angle valve job and some mild porting. Unshrouding helps flow because the combustion chamber walls won't block the flow if they are ground back a little. Note that the D16, like the B16A and the B18C5, has an open combustion chamber with no quench areas. This is not as good as the full quench combustion chambers found on the B18C and H22. Quench helps reduce detonation and ensures a more even burn of fuel.

factory. Sometimes even CNC machines are used to rough port heads for popular models of engine. Even though it is highly beneficial for power production, heads are usually not ported by the factory, as it requires too much labor-intensive handwork by a highly skilled machinist, which makes it very expensive.

The Extrude Hone process can also be used to port cylinder heads. In Extrude Honing, a thick putty-like slurry full of abrasives is pumped through the head, enlarging the ports just as the natural flow would want them enlarged. Lisa Kubo has used Extrude Honing in her 9-second, "world's fastest unibody" Honda rocket.

Limitations

Porting, as with most high performance modifications, has its limitations. It is possible to make the ports too big. An amateur cylinder head tuner armed with an air-powered die grinder can simply hog the ports out, making them as big as the Holland tunnel. Big ports can flow big numbers but big flow numbers alone will not make big horsepower. Larger ports have a lower flow velocity given the same flow demand. An air column of a given mass at a lower velocity has less inertia, and less potential energy, possibly negating any ram charging effect. The ram effect is critical for obtaining complete cylinder filling at low rpm and is just as critical for maximizing volumetric efficiency at high rpm. Incomplete cylinder filling at low rpm causes an engine to have poor bottom-end power and throttle response. Symptoms of an engine that has been ported too much include a soggy bottom end that only makes horsepower in a narrow, few hundred rpm range at top end, usually with a rough lumpy idle.

In carbureted or throttle body

DPR is in the process of modifying this combustion chamber from a H22 Prelude motor. The squared-off shape means that it has a quench style combustion chamber. The piston squishes the air/fuel mix away from the ends of the combustion chamber in a quench chamber, promoting a turbulent, even-burning mixture that is less likely to detonate. This head has the gasket area marked so the head porter knows how much he can unshroud the valves before intruding on the head gasket. A rough valve job and some porting in the bowl area have already been done.

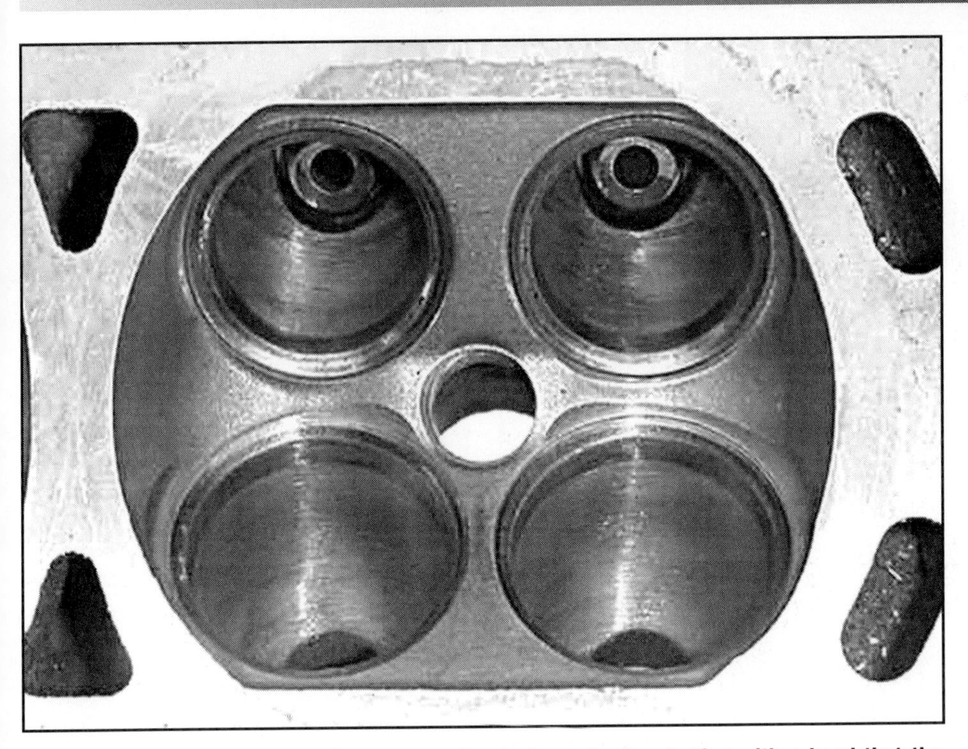

DPR works a B18C head for a low-compression turbo motor by starting with a head that the intake and exhaust ports are ported but the chamber is stock. This is a quench-style chamber.

Areas where material can be removed from the cylinder head to reduce the compression ratio and to reduce shrouding of the valves are marked out and ground away with a die grinder, as shown in process here.

injected motors, oversize ports with low velocity and stagnant flow can cause poor fuel atomization with its attendant bogs and stumbles. Typically, an engine with overly big ports, a high performance camshaft, and a carburetor will barely run at low rpm. Going too big in the ports can also mechanically weaken your head to the point where it flexes, blowing head gaskets frequently or even cracking. The main trick to effective head porting is making the "straw" big enough to feed your thirst but not making it so big that you can't suck hard enough to bring the milkshake to your mouth!

Truly effective porting is an arcane combination of craftsmanship, artwork and science. There are no hard and fast engineering rules that can be applied to all cylinder heads. Every type of cylinder head will respond a little differently. The optimal porting may vary in the same engine depending on what cam profile, intake or exhaust manifold the engine will be equipped with. To truly figure out the optimal pattern for porting a particular cylinder head, a head porter must spend many hours on a flow bench to find out what tricks will produce the best flow numbers, a process done mostly by trial and error. A flow bench is a machine that can measure the airflow through a cylinder head. A good cylinder head man always tries to shape the port to get the maximum flow with the minimal amount of enlargement. The goal is to keep the velocity as high as possible, and the volume low, but some engines or applications need greater emphasis on flow and ultimately need more port volume than others. A good head tuner will also try to get equal flow with each port so each cylinder will get the same charge of air. Most good cylinder head tuners have their own closely guarded shaping secrets for finding the magical diametric combination of low port volume, high flow velocity and high overall flow. What we are describing here are some general methods that head tweakers use when they modify heads. Your local head ace may or may not employ all of these methods or agree with what we are describing here.

Porting Variations

As stated in the previous paragraphs, line of sight porting almost always works better than stock, however, it is not always the best possible way to port a head. There are some exceptions to the line of sight rule that are counter-intuitive and must be proven on a flow bench or dyno because they don't always work. Although not applicable to Honda engines, the following are some examples of how non-conventional porting can extract more power. Engines with big bores and short strokes with valves located close together often prefer to have the port shape biased toward the cylinder

After grinding, the whole chamber is deburred and polished.

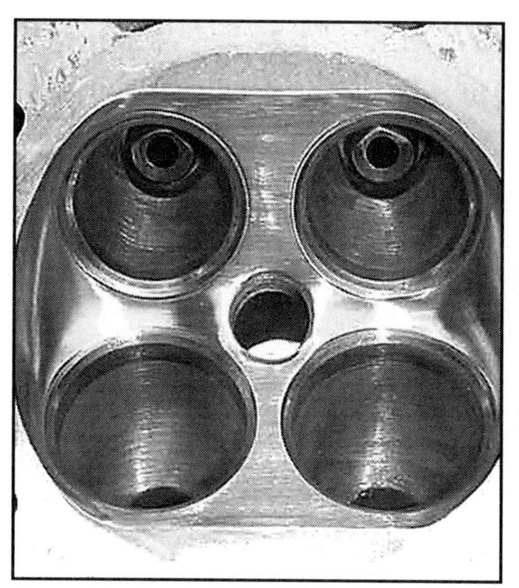

The final porting is completed and the valve job is finished.

walls. This prevents the intake charge from blowing right out the exhaust on overlap. Engines with a shallow-included valve angles often want the port's floor to have a hump near the inside radius of the valve seat to help point the airflow toward the bottom of the cylinder so the incoming charge won't get blown out the exhaust. Heads with valves near the periphery of the cylinder often want the ports biased towards the center of the bore to reduce the cylinder walls shrouding of the airflow out of the ports. These strategies cannot be predicted by using magic engineering formulas or rules of thumb. This is where dyno testing and hours on the flow bench will separate the men from the boys. This is where art can surpass science.

Other Factors Affecting Porting

How the engine is set up can influence porting. Turbocharged motors often prefer small, high velocity ports without a lot of overall port volume. Big ports on a turbo motor will often make turbo lag unmanageable. Supercharged and nitrous motors often prefer the opposite—large ports that flow a lot due to the increase of overall flow volume that nitrous and supercharging create. Street motors prefer high velocity, small volume ports much like a turbo motor, as do road-racing motors, which need broad, tractable powerbands. A naturally aspirated motor set up for drag racing and high rpm will like fairly large, free-flowing ports, sacrificing some bottom-end power for the top end, which is most important for drag racing.

VALVE JOBS

Another important aspect of headwork is the valve job. Believe it or not, over 50% of the flow gains generated by good headwork can be found in the valve job. Factory valve jobs often involve only cutting the seat, i.e., the 45-degree surface that the valves seal to. Sometimes there are additional rough cuts made to the port to ease the air's approach angle to the valve. This is why factory valve jobs are usually either called one- or two-angle. A one-angle job is done just on the seat surface; a two-angle job includes a seat surface plus a smoothing throat cut. In a cost-conscious, mass-production environment, there is simply no time or money to spend on small details like a multi-angle, precision valve job. If more flow is deemed necessary, it is usually cheaper for engineers to spec a slightly larger valve. Wide seating surfaces are more forgiving to mass production mismatches giving good sealing for the life of the car even with a significant misalignment of the seating surfaces. As valves and seats wear, they tend to sink lower into the seats, creating seating surface mismatches and valve shrouding. Imagine the valve having to lift itself out of a deep crater before it can start to flow, and you'll get the idea. Wide seating surfaces sometimes tend to last a little longer as the valves sink due to wear. Long life is an important factor on production engines.

As part of the valve job, some tuners remove the entire valve guide

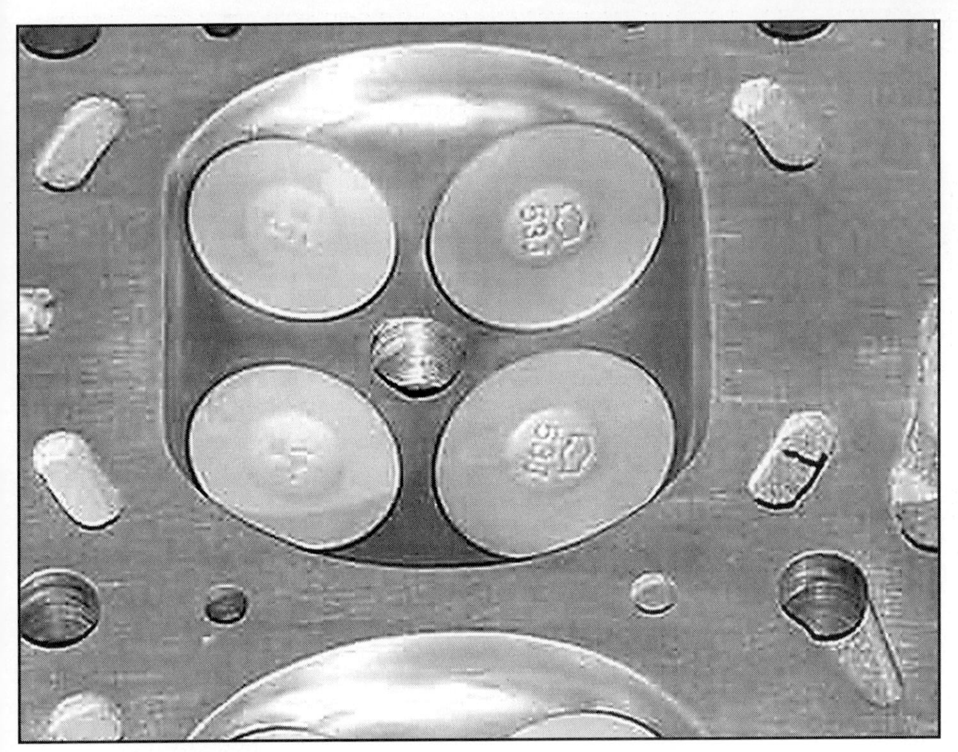

After a coat of ceramic thermal barrier coating, this head is ready for installation on a 9-second motor or a hot street motor.

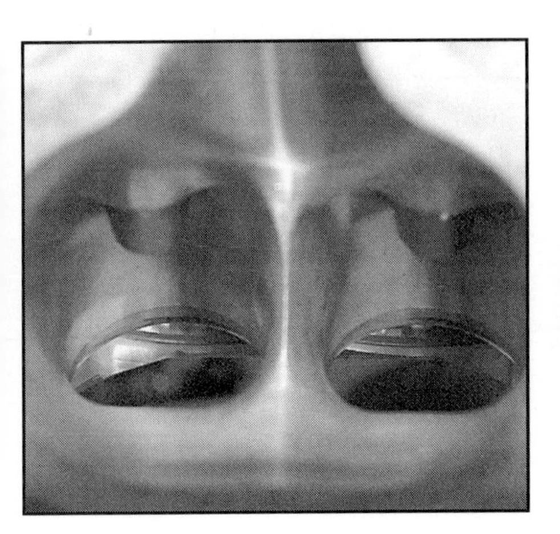

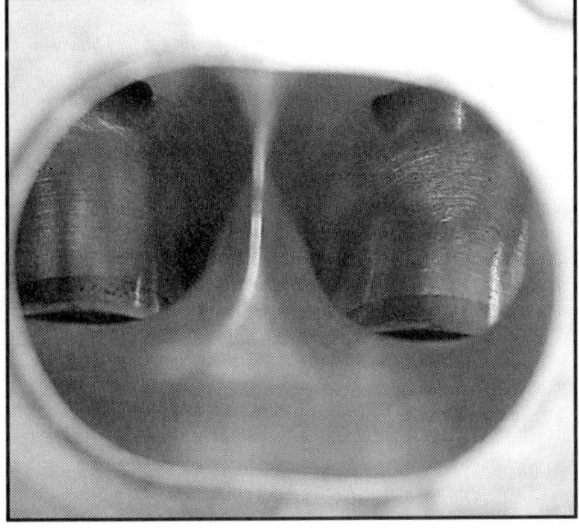

These worked-over Prelude H22 ports by Port Flow Design show raised port roofs, moved back port splitters and a smooth finish.

boss that extends down into the port, so the port will have the maximum cross-sectional area. Other tuners feel that rocker arm equipped motors like Hondas are better off running stock length valve guides as rocker arm motors tend to put more side load on the guides themselves and shortening the guides may result in additional wear.

Three Angle Valve Jobs

High performance valve jobs have three-angled cuts, one on each side of the valve seat. First there is the throat cut which is typically around 60–70 degrees. This helps ease the air's transition to the 45-degree seat cut.

The second cut is the 45-degree seat cut, which is the surface that the valve actually seals against. On a high performance valve job the seat cut is narrowed down to the minimum needed to seal the valve reliably and the seat's contact patch is matched accurately to the corresponding seat cut on the valve. On a mass production engine, wide production tolerances dictate that the seat cut ends up being much wider than actually needed. For a high performance valve job on a 4-valve motor, the seat cut is usually made to be 0.040" or so on the intake side and 0.050" wide on the exhaust valve. The exhaust is made slightly wider to have more contact area so the exhaust valve can be cooled better by conduction. With the valve seat width at minimum and having the seats matched to the valve, the incoming air or outgoing exhaust has less restriction due to the unshrouding effect the adjacent cuts produce. Simply put, as the valve moves away from the seat, more area for flow is exposed sooner.

The third and final cut is called the top cut. The top cut is generally about 30–20 degrees and is made immediately after the seat. The top cut also helps reduce valve shrouding of the airflow past the valve (or before as in the case of the exhaust valve) as the valve starts to lift off of the seat.

Five Angle and Radius Valve Jobs

Sometimes a head tuner will make a five-angle valve job, adding additional shallower angled cuts to make the entrance and exit to make the path to the seating surface even smoother to airflow. Some head tuners feel that very best valve jobs are called radius valve jobs. A radius valve job is a five-angle valve job where the two angles adjacent to the valve-seating surface are hand

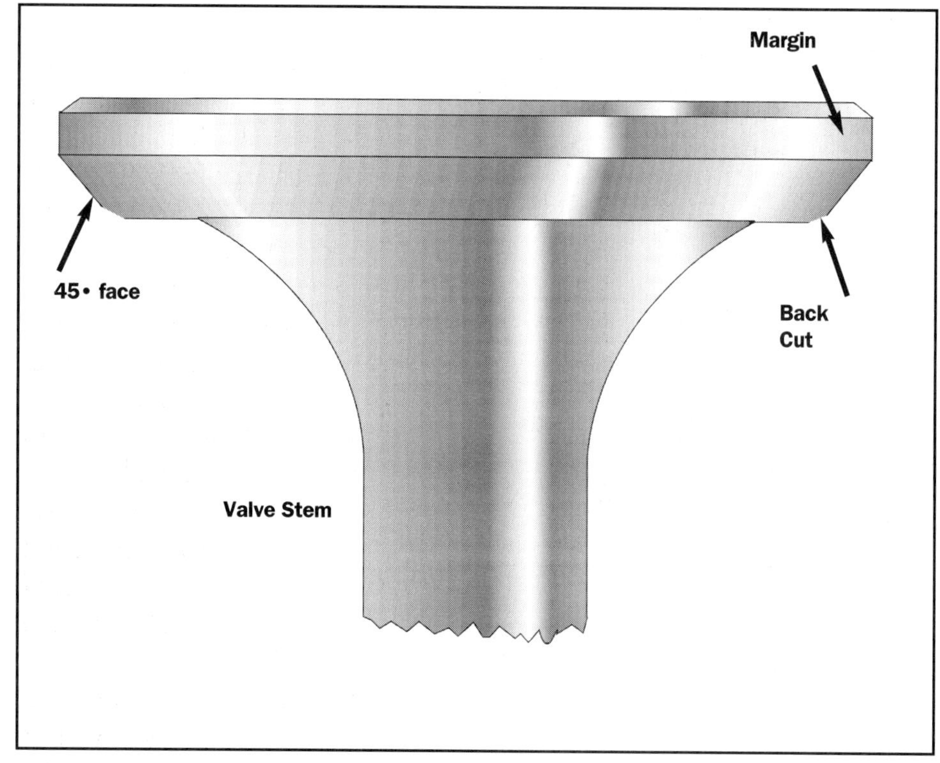

Valve terminology. Courtesy Dave Emanuel.

Here is what a 30-degree backcut looks like on a B18C valve. This backcut helps low lift flow.

blended together, into the port wall and combustion chamber for a totally smooth transition from seat to port to combustion chamber. These valve jobs are very labor intensive, hence expensive. Some head artisans firmly believe (and have proof from the flow bench) that this blending actually hurts flow, another example of art meeting science. Another trick to increasing flow is to have an approximately 30-degree angle backcut on the valves away from the seating surface. This also aids in unshrouding the valve to airflow as the valve starts to open, especially at low lifts.

Another trick is to reduce the diameter of the valve stem directly under the head of the valve. The valve stem is turned down a few millimeters, effectively shrinking the stem so it blocks less of the port. One should be careful not to overdo this, as overzealous grinding here can weaken the valve stem.

The head of the valve can be swirl polished to further aid flow. The swirl pattern reducing the boundary air layer thickness on the valve, much like the dimples on a golf ball, help its streamlining.

Valve Job Benefits

Near-opening and low valve-lift flow is critical in making lots of horsepower because the valves spend more dwell time at low lifts and close to opening than at full lift. In simple terms, the valve is only fully open once per intake or exhaust cycle but is near its seat twice as the valve has to both open and close! That is the primary reason why a simple valve job can make so much of a difference in flow and horsepower. Unlike most high performance mods where bottom-end performance is robbed to get better high-end power, valve jobs fall into the free horsepower category. Enlarging the ports can many times compromise low rpm performance as explained in the previous paragraphs. A high-performance valve job however, can increase power throughout the powerband with no sacrifices or drawbacks of any sort except cost and perhaps long-term durability, because the narrowed seats can tend to wear a little faster. Street high-performance valve jobs usually have a little wider seat cut for longer life although some head tuners dispute that narrow seats wear any quicker.

Some head artisans are of the opinion that valve jobs used for turbocharged, supercharged and nitrous applications need a little wider of a seat on the exhaust valve and should not have the exhaust valve guide shortened for better flow. This is because these types of motors generally run higher exhaust gas

This B18C head from Port Flow Design shows larger titanium valves, unshrouding and polishing. This head is from Jeremy Lookofsky's record-holding 10-second all motor CRX.

The best machine to do valve jobs is a Serdi. A Serdi makes the difficult job of doing a precision valve job easy. Here a technician at DPR does a valve job on a Serdi.

temperatures than equivalent naturally aspirated motors and need the additional contact area for improved valve cooling. Having the valve guide full length also contributes to improve valve cooling. Usually the seat width will be increased by 0.010" in these cases. Interestingly, many tuners do the same sort of valve job no matter how the engine will be fed.

Sometimes well-known modifiers of Honda heads install larger valves to increase flow. Usually these are about 1mm oversize and do not require different valve seats as the stock seats can be enlarged to accommodate a valve of this size. Some tuners offer stainless steel and even titanium valves as part of their Honda head package. Usually these valves have a modified profile for better flow. A valve with a flatter profile is installed in the intake side and a more tuliped valve installed in the exhaust side. Some head porters heavily modify the stock valves to have a reduced diameter stem and a tuliped backside just like a racing valve.

Valve Job Tools

Good tools are essential for a good valve job. Many head specialists use a drill-like fixture with stones of differing conicity to grind the various valve seat angles. A good valve job can be done like this but it can be labor intensive and difficult to maintain consistency from valve to valve. One sign of a good head specialist is the use of a Serdi Valve grinding machine. A Serdi is an expensive ($50,000) machine that makes precise, repeatable multi-angle valve jobs a snap with very little practice. Serdi machines pilot off of the valve guide hole with a floating power head to ensure precise alignment of the cut seat with the valve's axis. The depth of the angle cuts is also easily controlled. A Serdi machine makes the creation of a good

Another sign of a professional head porting shop is a Superflow flow bench. A flow bench measures the flow through a port and is an important tool for a head porter to use when figuring out the best porting configuration on a head.

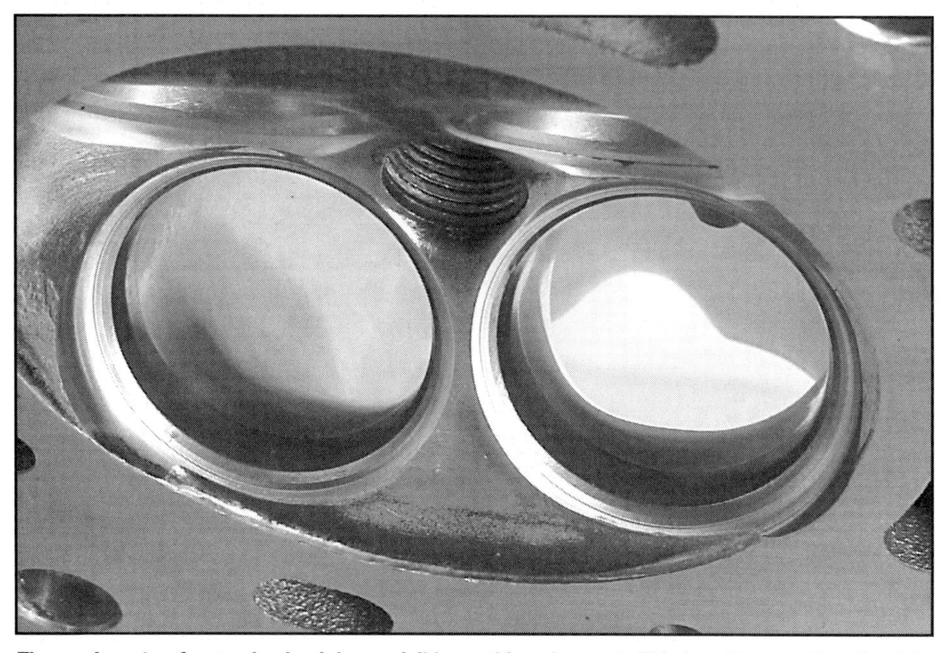

The angle cuts of a good valve job are visible on this valve seat. This is a three-angle valve job. Note the sharp edge on the combustion chamber. This can glow red-hot and cause detonation. This edge must be removed before the head is finished.

valve job easy.

The exact valve angles used by cylinder porters, like the geometry of the ports, are a closely guarded secret of many head artisans and racing teams. The angles and techniques listed here will vary from one cylinder head specialist to another.

COMBUSTION CHAMBERS

Some head specialists also modify the quench zones of the cylinder head's combustion chamber. The quench zones are the flat areas of the cylinder head where the piston comes in close proximity at TDC. The B18C, the '96-and-later D16 VTEC, and the H22 cylinder heads typically have 4 quench zones at the ends of the combustion chamber. Quench zones promote more complete burning and reduce the likelihood of detonation by increasing turbulence of the air/fuel mixture as the piston comes to TDC. It does this by squishing the air/fuel mixture towards the sparkplug and away from the end zones of the combustion chamber. This reduces the amount of air/fuel mixture near the ends of the combustion chamber, where it does not completely burn, thus being wasted by pushing or squishing it toward the centrally located sparkplug, where it can easily be ignited. When heads have additional quench area, they normally need less timing advance to make power. Thus, a skillful tuner can tune the engine to be farther from the detonation threshold, making the engine more reliable. Adding quench makes the motor more efficient, needing less ignition advance and producing more power due to more complete combustion.

Strangely enough, many of Honda's best motors have open circular combustion chambers, including the awesome B18C5 from the Integra Type R and the B16A from the potent Civic Si. These heads can be milled approximately 0.060" to turn them into a more effective and efficient pentroof chamber with quench zones. Milling will decrease combustion chamber volume and raise the compression ratio by about 3/4 of a point. You must CC the combustion chamber and calculate the true compression ratio of your engine to avoid problems associated with excessive compression like detonation on street motors. You must also purchase adjustable timing gears to restore proper cam timing, as milling will retard the

DPR, among others, does some fancy chamber work to increase quench in their high-compression All-Motor class cylinder heads. Here is a B18C head before undergoing the Stage 6 treatment.

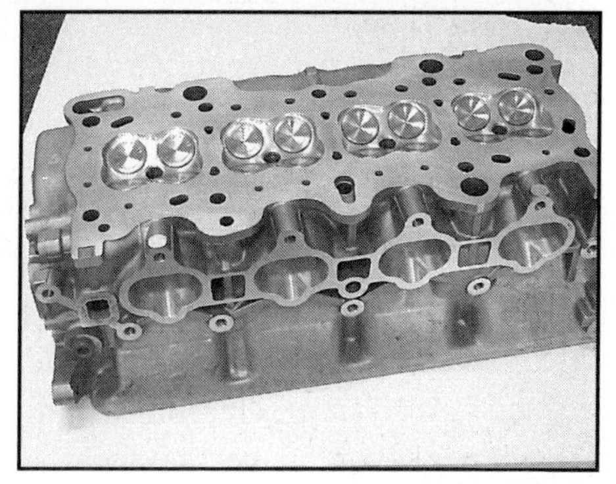

A complete Stage 6 head is a thing of beauty. Its improved quench makes the engine quite a bit more resistant to detonation and its smaller combustion volume allows the use of a flatter dome piston to get high compression, improving flame travel and port crossflow during overlap. This also helps improve engine efficiency. A Stage 6 head can add up to 15 hp to a stock B18C motor.

The quench areas between the valves are welded up, the head deck milled flat and the chambers reshaped.

cams by about 4 to 6 degrees.

CC'ing

CC'ing the cylinder head is done by placing a plate of Plexiglas with a small hole drilled in it over the combustion chamber, sealing it to the head deck with grease. Colored solvent is put into the combustion chamber through the hole using a calibrated burette. This allows accurate measurement of the chamber volume.

The piston dome or dish volume can be measured by backing the piston down the bore a known amount as measured with a dial indicator. The ring groove/piston crevasse area is filled with grease to seal it, the plate is placed over the top of the cylinder, and the resulting volume filled with solvent from a burette. The volume of the cylinder expressed by the diameter and depth of the piston is compared to the actual volume of solvent to determine the piston dome or dish volume. Now the compression ratio can be calculated. There are so many different variations of pistons and heads possible with Hondas that it is always a good idea to CC the piston dome/dish and head whenever you do a buildup to verify the true compression ratio.

Reshaping the Combustion Chamber

Adding quench to the cylinder can be taken one step further. The quench zones can be welded, milled and reshaped by hand to make them bigger, shaping the combustion chamber like a cloverleaf instead of the stock pentroof or circular open combustion chamber. This reduces the combustion chamber volume, increasing compression as well as making the quench zone more effective. This can also make the combustion chamber less likely to promote engine-damaging detonation, because the turbulent air/fuel mixture squished by the bigger quench zones burns more completely and smoothly. Cloverleafing the combustion chamber obviously makes a bigger difference on the open chamber heads.

The non-VTEC D-series SOHC motors, the pre-96 D16 VTEC motors, the B16A the B18C5 and the SOHC Accord motors lack quench zones in their combustion chambers and can greatly benefit from having the combustion chamber modified as such. Since it is obvious that some heads in the Honda line are better than others, we will detail which are the hot heads to swap in a later chapter.

CHECKING CLEARANCES

It must be stressed that milling and cloverleafing the head, in conjunction with pistons, adjustable timing gears and high performance cams, can have a significant impact on engine compression ratio and piston-to-valve clearance. Everything must be checked and confirmed before the engine is run or severe damage can result.

Piston to Valve

To confirm piston-to-valve clearance, a thick layer of modeling clay is placed on the piston dome and the engine is rotated two complete revolutions carefully by hand so any contact of piston and valves can be felt before damage occurs. The head is removed and the clay on the piston inspected where the valves have marked it. It is important to maintain at least 0.045" of clearance on the intake valves and 0.055" on the exhaust valves. This is the minimum clearance allowed and more is safer. You want at least 0.030" clearance between the piston and any part of the cylinder head also. When building an engine, unless you are doing a pure stock motor buildup, there are so many combinations of heads, pistons and cams possible on Hondas that you should always verify the piston-to-valve and piston-to-head clearance during the buildup.

The power gains produced by headwork can be significant. As we said at the beginning of this chapter, late model import 4-cylinder DOHC 4-valve motors as found on Hondas have pretty good modern heads to begin with and usually only experience 5–15 hp of power gain with even extensive headwork. This is much less than what can be found on the domestics, which generally have much poorer port layouts.

BOTTOM-END MODIFICATIONS 4

In this chapter we are going to get into the bottom end of the engine, which is very important for the production of big power numbers but is much ignored and misunderstood by all but the most serious car enthusiasts. This chapter will conclude our discussion on the inner workings of the motor. In the remaining chapters we will be discussing some of the peripherals that contribute to power such as turbos, superchargers and nitrous oxide systems.

As we talked about in our first chapter, the bottom end of an engine consists of the block, crank, rods, pistons, rings, bearings, oil pump and water pump. Its job is to mostly contain the explosive force of the combustion event, changing the outward expanding combustion cloud into smooth rotary motion. Since we got into much basic detail about how it does just that in the first chapter of this book, we won't do it again but we'll jump right into the specifics.

Most modifications done to the bottom end fall into two categories: One, stuff you do to strengthen the block so it can handle the rigors of increased power production, and two, stuff you can do to increase efficiency and power production for the engine's intended end application. In addition, turbocharged, supercharged and nitrous engines need some special preparation in the bottom end if they are to live for long at high levels of power.

Build It Strong to Last Long!

Honda motors are extremely durable engines, a fact proven on the streets every day. Honda's venerable B18, in both the non-V-Tech B and V-Tech C versions, along with the H22, are also bulletproof as is the B16A. Even the weaker little brother of these powerhouses, the Honda D16, is fairly strong with a lot of quality construction.

This is really not some pro import-biased BS either. Since the background of my youth was spent blowing up small-block Ford and Chevy motors, the strength and quality of import motors never ceases to amaze me. The small-block Ford in stock form suffers from rod bolts that are way too thin, weak connecting rods, flexible blocks and even more flexible head decks. The Ford 5.0 is also powerbound by a poor-flowing cylinder head. The venerable Chevy small block, the mainstay of domestic performance, doesn't have any significant weak areas, but it isn't suited for really high power output without major strength modifications. When I used to modify these motors, I had to massage or tweak almost every piece in the engine, from the rod bolts to the timing chain, if some decent power was to be reliably extracted. I spent many hours polishing rod beams, deburring and chasing holes in blocks with taps and shaving tons of metal in balancing. Lots of money was spent at machine shops having to align bore every motor and resize every rod as tolerances were so loose. Bolts had to be replaced with studs or

Honda engines have a very strong bottom end, far superior to most domestic blocks. For most mild modifications, the stock bottom end will do just fine.

high strength ARP versions of the stock hardware. Cranks and rods had to be shotpeened and nitrided. These domestic engines had many really stupid features also, like plastic timing chain gears that shed teeth with a harsh word and skinny oil pump drives that would snap, leaving you without oil pressure.

However, because these motors have been produced for over 40 years and are immensely popular, the performance aftermarket has developed reasonably priced solutions for most of this motor's weak points. But something seems wrong with the fact that you have to practically replace most of the engine with aftermarket components to get a stout high horsepower engine.

But enough domestic bashing. This is a Honda book, after all. When compared to these domestic blocks, the Honda engines basically stand up as far superior. For mild performance, which includes all bolt-on naturally aspirated modifications, low boost (10 psi) turbo kits, up to 100 hp of nitrous and supercharger kits, the bottom end of Honda motors can remain totally stock as long as you can avoid detonation. As I have said repeatedly in this book, detonation is the number one killer of performance motors.

The Honda D-series is a little less strong than its brothers, hampered by somewhat weak connecting rods. But their weakness is only relative to the amount of power being produced. For mild gains, they are just fine. They are still fairly strong as motors go with the same tough features. With the D-series, the revs should be kept below 7500 rpm and perhaps nitrous should be limited to a 50 hp shot or 7–8 lbs of boost from turbos or superchargers to be safe. Granted this is still pretty strong, but these limits should be remembered and adhered too for long-term durability.

BLUEPRINTING

Many enthusiasts don't really have a clear idea of what blueprinting is, believing there is some great science or voodoo involved. However, blueprinting is not some weird magic, it is simply the process of shifting the tolerances of the engine to the side of the tolerance that will give you the best power. Blueprinting usually makes a huge power difference in domestic engines because there is such a large variation in their clearances, but the improvement is less drastic in most import engines because their tolerances are controlled much more tightly during the manufacturing process.

Bearing Clearances

Many experts feel that the most important aspect of blueprinting is getting the bearing clearances to the loose side of spec to reduce friction. Most experts also feel that getting the cylinder piston to wall and ring end gap clearances to the tight end of the spec to maintain good ring seal and reduce blowby is important. Of course other experienced engine builders will have differing opinions on this but the important thing is to make sure that the clearances are all equal and controlled.

My personal recommendation is that you should maintain the minimum clearances on the pistons, crank and rod bearings for longer life and less windage-loss (windage causes oil to sling off), which will provide better ring sealing and maintain bearing oil pressure. On an SCCA showroom stock-class motor where no mods are allowed, I might run bottom-end bearing clearances a little looser, toward the middle or larger side of the clearance spec to reduce viscous losses, but everything else, like the pistons, gets clearances toward the tighter side of the spec.

Blueprinting bearing clearances on import motors is easy because they feature select-fit bearings. These are bearings available in several different thicknesses, so the exact clearance can be maintained for each journal, even if there is variation in the machining from journal to journal. The block, rods and crank are stamped with a code for what bearing size is used for each journal, and you can use a chart in the factory service manual to decode the stampings. If you want to tighten or widen clearances, you can go up or down a size from the encoded bearing size.

Cranks

Because most Honda cranks are so tough, it is rare that they have to be turned undersize, when the engine is gone through. This is a common rebuilding practice on domestic motors. If the engine has some miles on it, it is usually enough just to lightly polish the journals and use the next tighter sized bearing on the service manual chart.

Piston-to-Wall Clearances

Another important aspect to blueprinting is to get all of the piston-to-wall clearances for each cylinder equal. This is done by measuring each piston and having the machine shop match each piston to its bore with the exact same piston-to-wall clearance. Good machine shops can do this easily by honing each bore for a precise fit. Like the bearings, most import engines have select-fit pistons with the size used stamped on the block in a code that can be decoded with the factory service manual. If you are replacing the factory pistons, you can specify a piston grade a little bigger and hone the cylinder to get the fit. This gives you the benefit of a fresh bore surface, because you never want to put new rings in a used bore. They will not break in properly and will not seal. With aftermarket racing pistons, you do not have this select-fit option so the machine shop must match the bores to the pistons, which is not a big deal for a high-quality machine shop. The piston-to-wall clearances should all match to within 0.0001"– 0.0002" of each other. With aftermarket pistons, you are usually going for a significant overbore to get some more displacement. Most import engines can be bored 0.040" or one millimeter oversize with no problem.

Equal CC Volume

Making sure that the combustion chamber volumes of every cylinder are equal is a blueprinting job given the cylinder head porter. Usually a good head person will try to get all the chamber volumes to be within 0.5cc of each other so all the cylinders will have pretty close to the same compression ratio.

Balancing

Every engine should be balanced. By balancing the engines internal parts, vibratory stress on the engine is greatly reduced. An out of balance condition of a few grams can result in an unwanted load of many pounds in the wrong place. If these imbalance loads are added up, they can result in quite a bit more stress on an engine's insides. Balancing means dynamically balancing the crank, equalizing the weights of the pistons and the big and small ends of the connecting rods. This is accomplished by drilling holes or grinding the counterweights of the crank, sanding the rod caps and small end wristpin boss on a belt sander and carefully removing weight from around the pin boss and the underside of the dome on pistons. With pistons you have to be careful not to make the dome thinner that 0.200" or to weaken the pin boss. On domestic engines, I have found the weight of parts to vary by several grams, sometimes more than 10 grams, making balancing very critical. Lots of metal had to be removed. On tight tolerance Honda engines, however, the weight variation is usually less than a gram or two. Now you know yet another reason why I love these engines so much!

PISTONS

For extracting big amounts of power like you would if you were building a Quick Class drag car or even a crazy street car, more severe part replacements are needed for reliability and durability. Both import and domestic stock motors usually have cast aluminum pistons. Imports usually have high pressure cast pistons, which are superior to the more common low-pressure cast used by many domestics. High-pressure castings have a denser grain structure and fewer metallic inclusions to form weak points in their structure. Cast pistons are soft, which wears well with the harder cylinder walls. They also have limited expansion, which allows for tighter piston-to-wall clearances (around 0.0005"-0.003"), which helps emissions by reducing crevice volume, reducing oil consumption, and increasing ring durability.

The disadvantage of cast pistons is that they are relatively fragile, even the tougher high pressure cast ones. Cast pistons are brittle and tend to

This is a custom forged JE high-compression piston. The tall dome takes up volume when the piston is at top dead center, increasing the compression ratio. Unless ordered otherwise, most JE pistons are of the high-silicone variety.

This is a JE turbo low-compression piston. For turbo, supercharged and big nitrous applications, you generally want to run a lower compression ratio. Note that this low-compression piston comes with a dish instead of a dome. The added volume of the dome lowers the compression ratio. A properly designed dished piston still has raised edges to take advantage of any quench in the combustion chamber.

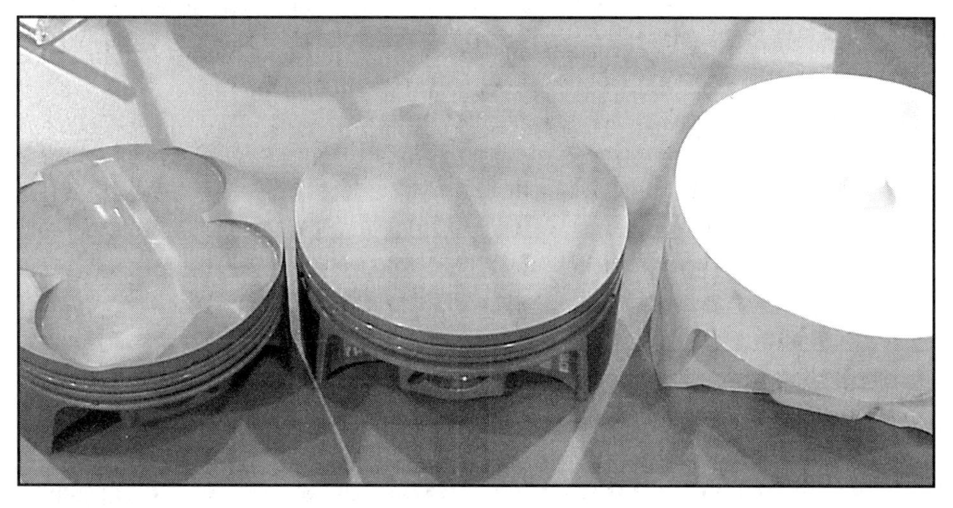

Here is how a forged custom piston is made, starting from the left. A blank forging is pounded out of a billet of aluminum by smashing it at ultra high pressures into a die. A rough-machined blank is CNC machined close to the desired bore diameter, the pin bore and boss is machined out and the ring grooves are tuned. Finally, the piston skirts are ground to the final diameter and skirt cam contour. The pin bore is honed to final diameter and the valve reliefs and other dome features are milled into the top of the piston.

crack if there is detonation or if they are subject to a lot of load. To avoid these problems, you must use a stronger piston if you are building a high-powered motor with either a high-boost turbo, a big nitrous kit or a supercharger. In this case, you will want to run a forged piston.

Forged

A forged piston is made of an aluminum blank that is smashed into a die at extremely high pressures. This forging process aligns the grain of the metal around the piston for maximum strength. As an example, metal has a grain just like wood, and you want the forces to be with the grain instead of against it. The pressures needed to mash a cold piece of aluminum also refines the grain, making it tighter and finer, increasing the strength and hardness of the aluminum by cold-working it. I will not get into the metallurgy here as it is beyond the scope of this chapter, but let me say that forging makes the metal stronger and more ductile.

Types—Forged pistons come in two different varieties: high silicone and low silicone. High-silicone pistons are more brittle but expand less under heat. They are still much tougher than the stock cast piston but are not the toughest, being more brittle than the low-silicone versions. High-silicone pistons can run a much tighter piston-to-wall clearance, usually about 0.002"-0.003", than low-silicone pistons, which makes them the piston of choice for endurance-type racing, and super-hot street cars. The tighter piston-to-wall thickness is easier on rings and cylinder walls as there is less wear-inducing rattling and banging going on, and they are more quiet for this same reason.

Low-silicone pistons are the toughest. Their alloy is more ductile, making them the piston to use for extreme abuse like drag racing with big boost motors. Their major drawback is that they expand more under heat so they must run a bigger piston-to-wall clearance of about 0.003"-0.006". This makes a clattering sound when cold and sometimes even when the engine is hot. An engine with low-silicone pistons often sounds like a diesel, especially when the engine is cold. The wider clearances can increase the amount of wear, because the cylinder walls and rings are battered when the piston rocks around in the bore. That is why low-silicone pistons are often only used for short duration events like drag racing.

Forged pistons must usually be custom ordered, but JE, Aries, Wiseco, Cosworth and Ross carry them in stock for most Honda motors.

JE has some interesting new technology. They can scan and digitize your engine's combustion chamber and use this data to come up with very specific dome contours that match your combustion chamber exactly. This can be a great advantage when designing a piston for an ultra high-compression, all motor class engine so the dome does not interfere with flame travel. Using this method, quench can be built into pistons for open chamber heads like the B16A or B18 C-5 heads without milling or welding the combustion chambers

Rings

For guidelines when ordering pistons, keep these things in mind. If you are going turbo or other forced or chemical induction, it is a good idea to run the first compression ring about 0.250" below the crown of the piston. This gives your ringland the strength you need to hold up to extreme use. You do not want your piston dome to be less than 0.200" thick. I believe that the best rings to use are late-model OEM Japanese rings for the correct bore size. Stock rings from late-model Honda motors are thin to reduce their mass and inertia to help prevent high rpm flutter. This is when the ring floats in its groove, skips and loses sealing when the piston changes direction at TDC very rapidly at high rpm. A light ring is less likely to do this. Stock Honda rings are also low tension to reduce friction and further reduce the likelihood of float. These rings are also chrome faced for long wear with the correct cylinder wall finish. I feel that it is best to stick with the stock rings unless you are using a custom bore size.

The holes in the top of this high-compression D16 piston are gas ports. These let combustion pressure down behind the #1 compression ring, pushing it into the cylinder walls, helping high rpm seal.

Here are gas ports on a B18C piston crown.

Some experienced engine builders disagree with this idea. A popular aftermarket choice among many engine builders is the Speed Pro gapless piston ring. This has a special second compression ring that has a special way of butting the ends of the

Here is the gas trap groove on a Venolia custom piston.

Here is a very large gas trap groove on a JE piston. The many grooves on the piston before the top compression ring is supposed to act like a buffer to reduce detonation stress on the top compression ring.

ring together so there is no gap. This ring gives excellent static leak-down numbers. I still prefer to use the high-tech Honda stock rings because of their high quality and construction features.

Some interesting features that can be tried with a custom forged piston are gas porting and gas trapping. A gas port is a series of small holes drilled into the dome of the piston down to behind the first compression ring groove. Combustion gas pressure is tapped and used to push the first compression ring hard into the cylinder walls to help improve ring seal, especially at high rpm. This is something that should only be considered for drag racing or other short duration forms of Motorsports because it accelerates ring wear.

Gas trapping is machining another groove between the first and second compression ring. This groove helps to reduce the gas pressure the second ring sees by serving as a reservoir for any gas leaking past the first ring. This improves ring seal at high rpm. This trick comes from Indy and F-1 cars as well as motorcycles. Although there are no drawbacks to this, it is still not a common practice on American aftermarket custom pistons, except for possibly Cosworth.

Dome Configuration and Compression Ratio

The piston dome configuration is also a big factor in finalizing the compression ratio. The compression ratio is a measure of how much the engine compresses the air/fuel mixture on the compression stroke. The ratio is the volume of the cylinder when the engine is at TDC over the volume of the cylinder when the motor is at BDC. A mild naturally aspirated motor where you have no control over the mapping of the ECU can usually run a compression ratio of 10:1 no problem on 92-octane pump gas. If you can program the engine's ECU or are running a user-programmable, stand-alone ECU, compression ratios of up to 11.5:1 on 92-octane pump gas are possible. All-out, race only, naturally aspirated motors using only racing fuel, like those used in All Motor 8 class drag racing or road racing, run compression ratios of 13:1 or even as high as 15:1. High compression ratios increase an engine's thermal efficiency greatly and slightly increase the volumetric efficiency. For every point of compression increase, you get approximately 3–4% more power. This is increased power across the board at all rpm ranges. The increase in thermal efficiency from raising the compression ratio also improves the Brake Specific Fuel Consumption or BSFC. This is a measure of an engine's thermal efficiency measured in pounds of fuel per horsepower per hour. The proportional gain in power falls off sharply at compression ratios above 14:1 and it is almost impossible to get a detonation-free burn at higher compression ratios than this, no matter what the fuel.

High-compression pistons normally have a dome; the dome takes up cylinder volume at TDC to raise the compression ratio. To raise the compression ratio to what you desire, the dome volume must be stated to

This cutaway of a JUN prepped B18C shows the piston dish needed in a low-compression turbo motor.

the piston maker so they can design the appropriate dome. A good idea is to keep the dome as flat as possible as too high of a dome can hamper flame travel, chamber scavenging on overlap and cause unstable combustion, which can cause detonation or give a loss of power. The ideal piston dome configuration in a perfect world is a flat top. This gives best combustion and minimal surface to volume for good thermal efficiency but on Honda engines, especially all-out naturally aspirated racing engines, the dome is often needed to get the desired compression ratio.

Some engine builders weld up the quench zones on the cylinder heads so higher compression can be achieved with a flat-top or low-dome piston. This is a good idea but excessive chamber welding can cause valve shrouding and reduced port flow so there are limitations to this technique. Any drastic chamber welding should be proved out on a flow bench to see if it reduces flow. It is not worth it to reduce flow to get a perfectly flat piston; a slight dome on the piston is preferable to sacrificing flow.

Forced induction engines, mainly supercharged, turbocharged, and nitrous kits over 100 hp, require lower compression to avoid detonation, especially with 92 octane pump gas. Lowering the compression allows for more boost pressure without having to retard the spark too much. Too much spark retard can result in excessive heating of the combustion chamber, valves and turbocharger. At a certain point due to this heating, retarding the spark can even contribute to detonation. Lowering the engine's compression ratio allows the tuner to have enough spark advance to avoid the viscous circle of retarding the timing to avoid detonation, only to have the increased combustion temperature contribute to detonation.

Some famous engine builders like JG disagree with this and advocate high compression ratios for race-fuel-only turbo motors (11:1), but conventional wisdom indicates that a low compression of about 8.5:1 or lower is desirable for a healthy, turbocharged, motor. This is usually achieved best with a dished piston. A dished piston is just how it sounds. The compression ratio is lowered by carving a dish into the piston to increase combustion volume. The configuration of the dish should be so the top edges of the piston around the circumference of the top, the areas that are not dished, still come into close proximity of the combustion chamber's quench areas. We discussed what quench areas are on pages 52–53. A dish piston configured like this still allows the quench areas to function correctly. These pistons are called "reverse deflector pistons," and resist the formation of detonation because of the turbulence induced by the quench areas, and end gas elimination from the periphery of the cylinder.

Piston Pin

The piston pin is the steel pin that connects the piston to the connecting rod. The very best piston pins are made of a high-strength, wear-resistant steel like H11. The strength of H11 allows the pin to be thinner and lighter for the same strength when compared to regular steel pins. Having good piston pins is important because pin failure is always catastrophic, nearly always destroying the whole engine. The best pins are taper-walled. Taper-walled piston pins are thick in the center where they pass through the rod bushing but taper out

Adam Saruwatari's radical NSX sports a mix of fairly conventional and radical technology, a forged, low-compression, iron-coated (necessary for running against the NSX's carbon fiber reinforced cylinder liners) Wiesco piston with a lightweight steel piston pin and a billet Sanz connecting rod. Note the bronze bushing in the small end of the rod for the full-floating piston pin.

thinner at the ends. This puts the metal where it's needed most and eliminates a stress riser in the pin as well as excess weight.

Floating Pins—Import engines have floating piston pins. Floating pins are ones that ride in bronze bushings in the small end of the connecting rod. Domestic engines use piston pins that have a press interference fit to the rod. The easy-spinning floating pins are better at high rpm because they are less likely to fail due to galling in the piston's pin boss and have less friction than the pressed pins. The issue with floating pins is that they must be securely retained in the piston or they can walk out and destroy an engine's cylinder walls. Floating pins are usually retained in one of three ways.

The first is called a Cosworth clip because Cosworth racing engines were the first to use them. A Cosworth clip is a simple circle of round spring wire that is retained by a circular groove in the piston. The Cosworth clip is easy to use and retains well. A Cosworth clip must be used with a beveled piston pin. The beveled piston pin pushes the clip into its groove tighter with load. A conventional pin can cause a Cosworth clip to pop out with disastrous consequences. Cosworth clips are common on Cosworth pistons (naturally) and pistons from Japanese tuners like JUN and Tomei. Many stock import motors use clips like this also. Be careful not to mix up the type of piston pins with these clips!

The next is an internal snap ring. These are just heavy-duty versions of your regular snap rings. They are strong and easy to use with snap-ring pliers. They can be used with flat-ended piston pins. Some pistons use double snap rings to be extra secure. A caution with snap rings is to make sure the side of the snap ring with the sharp edge faces outward. If the snap ring is placed facing inward, the rounded edge can cause the snap ring to cam out. Snap rings are popular with American aftermarket pistons.

Finally there is the spiral lock and double spiral lock. A spiral lock is a circular clip made of many passes of spring steel. Spiral locks are the most secure piston pin retainers. However, they are destroyed in the removal process and are a little difficult to assemble, as they must be wound into the groove. Some pistons have the extra secure double spiral lock. Spiral locks are popular on American aftermarket pistons.

Any of these methods is superior to the fixed press-in piston pin.

CONNECTING RODS

Connecting rods are perhaps the most important part of the bottom end. The connecting rods must be strong to withstand the incredible stress and load, which rises exponentially with rpm. Because of their dynamically imposed loading, rods must be light as well as strong. Rods must also be bulletproof because their failure will claim the entire motor. Stock Honda rods are usually very strong compared to their domestic counterparts. Honda rods are forged. Domestics cast a lot of their rods. Hondas have beefy rod bolts and spot-faced shoulders. Domestic rods often have spindly cap bolts and thin, stress-riser-ridden broached shoulders. Broaching thins the rod in this critical area and causes at least two large stress risers in this critical area. In case you did not know, sharp edges on parts cause stresses to concentrate there and the part usually breaks in these areas of stress concentration.

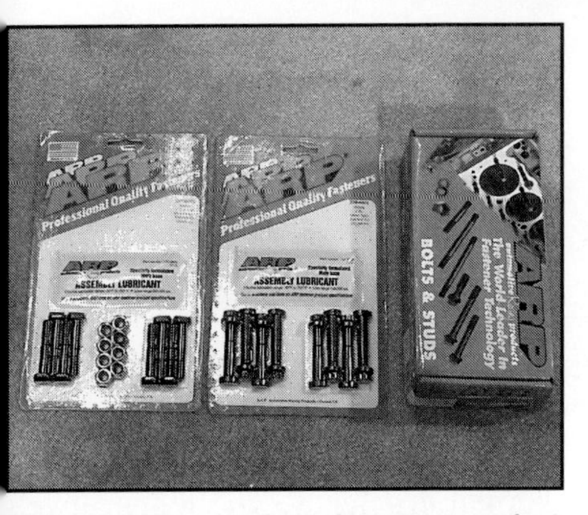

ARP rod bolts are cheap insurance against bolt failure. They are much stronger than the stock bolts and are inexpensive. Broken rod bolts are the most common source of rod failure.

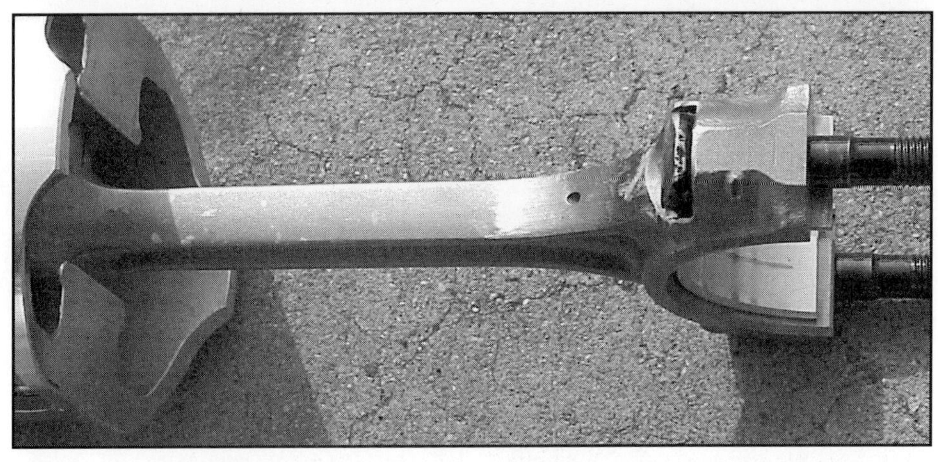

This is a D16 rod that has been prepped by polishing the parting line out of the rod beam, installing high strength ARP bolts and shotpeening. After this work, the rod almost looks as nice as a real racing rod.

The perforations in a postage stamp or a sharp fold in a piece of paper are simple examples of a stress riser.

Because of these features, unless you plan some crazy modifications requiring lots of boost, nitrous or insane rpm, you can get away with stock rods in your motor. Compared to other Hondas, the D-series engine has marginal rods. Although these rods are still strong enough for most bolt-on mods, the limitations to them can be reached when going for all-out mega power. These rods become marginal much past 200 hp and 7500 rpm. This means keep it to a 50 hp shot of nitrous, and under 8–10 psi of boost. For a low buck fix, the rods from an Integra motor will interchange and are much stronger.

Rod Prep

If your rods are to be used in a semi-serious street motor, they can be prepped by polishing the rod beams and shotpeening them. Polishing the beams removes the forging parting line, which is a major stress riser, and shotpeening improves the fatigue strength by about 100%. After shotpeening, the rods should be checked to see if they need to be resized and straightened as the force of shotpeening can distort them. Shotpeening is blasting the rod with steel balls of a controlled size at a controlled velocity; it is not sand blasting or bead blasting. Shotpeening cold-works the metal surface and refines the grain, sort of microforging the metal. This compressed outer layer strongly resists the formation of cracks. Small cracks can rapidly grow into larger ones. Shotpeening greatly improves the number of cycles near the yield point a part can take before failure. Don't let a machine shop tell you otherwise. It is not as important to shotpeen your rods as a domestic motor because many of the strong, popular import motors have shotpeened rods from the factory.

Rod Bolts

Rod bolts can also be replaced with high strength aftermarket bolts in many cases from SPS or ARP. They usually won't have the exact drop in application for your motor but SPS or ARP will have something close. The stock rods can be reamed slightly larger for the aftermarket bolts. For most applications, however, the stock bolts are adequate. If you feel you need to replace rod bolts because you have such a killer motor, then you should also be considering racing rods.

Racing Rods

If you are building a motor for real racing or a killer street machine, you may want to consider racing rods. Racing rods are made with superior metals such 4340 alloy steel. The rods are machined from solid billet, or in the case of Cunningham rods, forged 4340 blanks. Racing rods are usually completely polished to get rid of all sharp edges and tooling marks on their surface. Most racing rods are also shotpeened to further improved fatigue strength. The rod bolts are high strength, made of stuff like H11 tool steel with over 200,000 psi of tensile strength. Normally, racing rods are made to such close tolerances that they do not need to be balanced.

Racing rods are good for boost levels above 15 psi and over 8000 rpm. There are basically two types of rod beam configuration, "H"-style and "I"-style. H-style rods are generally considered superior by engineer-type purists because they have a bigger section modulus (engineering term for a measure of stiffness relative to cross

The Crower 2000cc B18B stroker kit has the best of everything: forged high or low compression pistons, excellent billet connecting rods and billet crank.

section) in the direction of the rod's bending load. Carrillo and Eagle rods are good examples of H-style rods. However, there are a lot of excellent I-style rods available. These emulate the external cross section shape of most stock rods. Crower and Cunningham are examples of high quality I-style rods.

A rod failure always means the total destruction of the motor, so buying racing rods is never a bad investment for the serious engine builder. It is very rare for a racing rod to fail, even on an all-out racing motor. Usually, the only way one will fail is if a bearing spins and the rod goes metal to metal on the crank, or if the engine ingests a good chunk of water at high speeds, in which case the whole motor is usually destroyed anyway.

An interesting note is that since the tensile load on the rod goes up exponentially with the rpm, a screaming 10,500 rpm 280 hp N/A class motor puts more destructive tensile stress on the rod than a 500 hp turbocharged Quick Class car chugging at 8500 rpm. The rod can handle the compressive loads that turbo boost provides better than the bolt-stretching spinning that it takes to make power with a naturally aspirated engine. That is one of the neat reasons that near-stock motors can make a lot of power when boosted if they are properly tuned!

INSIDE STUFF

A commonly known but often forgotten fact is that the internal geometry of an engine can affect the engine's power delivery. Bore size, stroke and rod length all have a profound effect on how an engine behaves in its power delivery characteristics.

Rod Length Ratio

An area of tuning that is just now being used by Honda engine builders is the altering of the stroke-to-rod length ratio. The bigger the stroke-to-rod length ratio the more dwell time the piston has around TDC. This accomplishes several things. Since the piston is near TDC longer, the combustion event has a longer time to impinge upon the piston, allowing a better transfer of force to the piston, slightly improving the engine's thermal efficiency. The longer dwell time also gives more time to fill the cylinders during the intake stroke and more time to scavenge the cylinder during overlap, improving volumetric efficiency. With a short stroke and a long rod, the point where the crank pin to the rod angle reaches 90 degrees, otherwise known as the point of highest piston acceleration, is further down the bore. Thus the piston accelerates more gently away from TDC. Since the piston is accelerated more gradually away from TDC, there is less mechanical stress on the crank, rods, pistons and cylinder walls. Reduced rod angularity at the point of highest cylinder pressure (about 30 degrees after TDC on the power stroke for most engines) also reduces mechanical stress as the piston digs into the bore underside load less.

Higher rod ratios have less velocity in the intake ports and there is a lower demand for the ports to flow as well because there is more time available to fill and scavenge the cylinder. Conversely this can also mean stagnant flow at low rpm, which is not too good for low-rpm power production. You cannot get everything for nothing!

To increase the stroke-to-rod length ratio, some Honda tuners are running a longer connecting rod. Moving the piston pin up higher into the piston allows this. Some engine builders are even running deck plates to raise the engine's deck so a longer rod can be run. Some Honda motors have stroke-to-rod length ratios as low as the H22's 1.49:1, 1.7:1 or better. The high-revving B16A is considered good. The most highly developed 4-stroke engines in the world, such as Formula One and motorcycle engines, often have stroke-to-rod length ratios of over 2:1, and many tuners of production engines are attempting to emulate this.

Stroke-to-rod-length ratio can also affect how the engine responds to changes in the cam profile. With a longer rod, since the piston travels away from TDC slower, there is more time to fill the cylinder with fresh fuel and air. Because of this, the camshaft's overlap period can be reduced and more cylinder pressure can be built at lower rpms due to the reduced blowdown and reversion resulting from less overlap.

The NIRA all-motor champion CRX built by AEBS and driven by Rodger Sango is powered by a B18C that has a deck plate welded to the top of the block. The deck plate allows for a much longer rod to be used, maintaining a stroke-to-rod length ratio of over 1.7:1, even though the engine has been bored and stroked to 2100cc. Paulus Lee, AEBS's chief engineer, was the first to exploit the potential of a bigger stroke-to-rod length ratio.

Bore-to-Stroke Ratio

The bore-to-stroke ratio of an engine can also affect the engine's power characteristics. Oversquare engines, ones that have a bigger bore than stroke, have lower piston speeds and less internal stress at high rpm due to lower inertial loads. There is also more time to fill the cylinders because of the lower piston speed. Longer strokes with smaller bored engines, called undersquare, have more internal stress due to faster piston acceleration, higher piston speeds, which accelerates wear and can induce seal-killing ring flutter. Undersquare engines have higher, torque and low-end power producing intake port velocities to ensure more complete cylinder filling at low rpm. Honda automotive engines are undersquare.

Engine designers can get around things like engines being undersquare with short rods by designing around these things. Despite being undersquare with a low rod ratio, most Honda engines can still rev to the moon because their port configuration gives good airflow, even at high port air velocities, and the huge duration and lift of their high-speed VTEC cam lobes ensures good breathing at high rpm.

With a good bore-to-stroke ratio, stroke-to-rod length ratio and a modern, free flowing head, there is less need for a long overlap period as the piston has more dwell time in crankshaft degrees around TDC. Although the difference is literally milliseconds, this is a considerable amount of difference. Engines with shallower valve-included angles, like Hondas also tend to have more crossflow on overlap and also can breathe well without long overlap periods. In general, the head's flow characteristics and the engine's bottom-end configuration have a lot to do with what kind of cam specs the engine will ultimately like.

Proof of this theory lies in practice. High performance motorcycles (with a few notable exceptions like the torquey Supertwin racing class bikes), and purpose-built racing engines, like the ones found in Indy and F-1, are oversquare for cylinder filling at high rpm and reducing piston speed reasons. These motors are usually very oversquare, have extremely free-flowing heads and have big stroke-to-rod length ratios in the 2:1 or higher range. These motors run very little cam overlap, and sometimes, as in the case of Honda inline 4s found in the CBR series of bikes, zero overlap! A less accessible, highly developed, 4-stroke are the ones found in Formula One cars whose cams have

RS Motor Works has deck plates for several popular Honda motors so a longer rod can be used and better rod ratio obtained.

surprisingly little overlap considering the duration necessary to ensure breathing at 17,000 rpm.

An engine may be limited in power production by a restrictive head design or, by rpm limits due to the weakness of its parts, or by the poor sealing ability of the rings due to flutter because of piston speed. To compensate, a tuner can destroke the motor with a shorter stroke crank, use longer rods and sleeve the block for a bigger bore to help with high rpm breathing and durability.

I feel that to further exploit their potential, Honda automotive engine tuners should look toward high tech ring packages that allow the piston pin to be placed into the oil ring groove area and even add deck height so longer rods can be run. Longer rods and a more oversquare bore and stroke combination work in motorcycle and unlimited racing engines. These ideas can probably be carried over to modifying production-based Honda engines in the future.

THE CRANKSHAFT

The crankshaft is one of the strongest parts of a Honda engine. Most Honda motors have forged crankshafts, as opposed to the cast cranks found in most domestics. Like forged pistons, forged cranks have superior grain orientation and grain refinement from the forging process. On popular Honda motors, the crank is one of the parts that need the least amount of work to ready it for even the most extreme use. Stock-prepped Honda B- and H-series cranks have supported more than 600 turbocharged horsepower without failure. These stock cranks are so strong that they have held up to the rigors of SCCA GT class road racing.

Prep

The crankshaft really only needs to have its pressed-in oil galley plugs drilled out and replaced with screw-in Allen-type plugs that are secured with Loctite™ and staked into position to ensure they will not come loose. While the plugs are out, the oil galleys are thoroughly cleaned with a brush and solvent. It is amazing how much nasty gunk and debris get clogged into the plugged blind holes of the crank. The oil holes on the journals should be lightly chamfered with a die grinder (which is done stock usually) and you should get a good dynamic balancing job. The journals can be given a light polish with microfinishing paper, but that's about all you need to do. Major modifications are not necessary even for wildly boosted motors! Many a 500-hp plus motor has been assembled with no crank prep work at all. But if you want to go a step further, read on.

Grinding—If you're really picky and want to cover all bases, you can take crank prep a step further by grinding off all forging parting lines and bull-nosing (rounding off) the counterweights. This reduces stress risers by eliminating sharp edges where stress and cracks can concentrate.

Lightening—All-out drag racing NA motor cars, where every ounce of power is critical, can benefit from knife edging and lightening the crank. This is where the counterweights are given an aerodynamic profile with a sharp leading edge to reduce windage losses when the crank spins through the oil-filled mist created in the crankcase by bearing sling-off at high rpm. Lightening the crank reduces the power required to spin the motor up to speed, freeing up power to drive the wheels. However, removing large

Here is an example of an Allen pipe plug installed in the oil galley passage of a modified stroker B18C and a D16. Threaded plugs will never unintentionally come out to blow an engine's oil pressure and can be easily removed for cleaning when the engine is freshened up between races.

Here is a forged R&D Racing stroker crank for the D16 motor. R&D can get as much as 2100cc from this motor! Note the knife-edged counterweights that reduce weight and windage losses.

amounts of material from the crank counterweights can create an underbalanced situation where the counterweights do not cancel out the weight of the rod and piston. This can create vibratory stress on the motor. Because of this, excessive lightening should be avoided on road racing and street motors where long-term durability is important.

Shotpeening & Nitriding—Other details that can be done that are nice but not absolutely necessary, even for the most powerful motors, are shotpeening the crank journal fillets to improve fatigue strength and nitriding or cryotreating the crank. Shotpeening creates a fine-grained compressed layer in the critical fillets area of the crank. This helps prevent the formation of cracks. After shotpeening, the crank should be checked for straightness and corrected if needed. As a note, many import cranks, like Nissans, are pre-shotpeened from the factory. Look for the telltale pebbly surface texture on non-journal machined areas.

Nitriding and cryotreating improve the abrasion resistance of the journal's surface. Nitriding surface-hardens the journals and makes them wear resistant. Nitriding also slightly improves the fatigue strength. Cryotreating also improves fatigue strength and relieves internal stress. These details are nice, but as I mentioned, many a 500 hp-plus import motor has been built without them.

Unlike softer domestic cranks, import cranks rarely need to have the journals ground undersized and have thicker bearings installed. All the journals really need is a light micropolish.

Stroker Cranks

For some motors, mainly the Honda B18B and C, the B16A stroker cranks are available. Crower and JUN are two of the main suppliers of these cranks. Stroker cranks have longer strokes to gain power-producing displacement. Stroker cranks are usually formed from a solid chunk of alloy steel such as 4340 and CNC machined to final dimensions. These stroker (or any billet crank for that matter) is bulletproof as well as super expensive. Billet cranks normally don't need any prep work, but the perfectionist will still want to brush-clean oil passages, check the balance (normally it's perfect on a billet crank), micropolish the journals, and deburr.

The Crower and JUN stroker kits include rods and pistons, as a long stroke necessitates a higher compression height, which means that

Here is the beautifully crafted and expensive JUN 2000cc stroker kit for the B18C. JUN also makes a great 1800cc stroker kit for the B16A,

R&D's H22 stroker kit features a modified F22 Accord crank for mild buildups of up to 2500cc. With a billet crank, as much as 2700cc is possible! Imagine what a killer street motor that would make.

the pin location in the piston and/or the connecting rod length must be changed so the pistons don't poke above the block's deck with the longer stroke crank. The bigger stroke also greatly increases the compression ratio so changes to the piston's dish or dome are needed to maintain the same compression. These kits are expensive, but remember there is no replacement for displacement! Power gained from a displacement increase is across the board but more gain is usually found on the bottom end of the rev range.

R&D Racing offers super stroker and big bore kits for the D, B and H series Honda engines, getting huge displacement increases in the order of half a liter or more from some of these engines. These kits are the secret super weapon of many of the leading All Motor class import draggers. A detailed look at R&D's motors will be featured in a later chapter of this book.

Gary Kubo, husband and engine builder for Lisa Kubo, who currently drives the world's fastest Quick Class Honda, has done some innovative work with destroking the B18C, shortening its stroke to improve the bore-to-stroke and stroke-to-rod length ratios. Gary reports that there is more power to be found this way due to improved breathing and mechanical efficiency despite losing some displacement.

THE BLOCK

Although Honda blocks are basically a very sound design, there are a few weaknesses that need to be addressed when going for hard-core power. Although these block mods are not needed for hot street motors, not even very hot street motors, they have to be done for the engine to live under racing conditions, be it very high rpm All Motor class cars or mega-boosted Quick Class cars. Although the area of block supporting the crank is nearly bulletproof, with either a full or partial girdle and beefy main bearing caps, the cylinder walls and deck are the Achilles heel of the mega power Honda. The popular B-series and D-series motors have what is called an open deck block construction. This means the cylinder is free standing in the block's water jacket with no support except at the bottom of the cylinder. The Honda factory has two

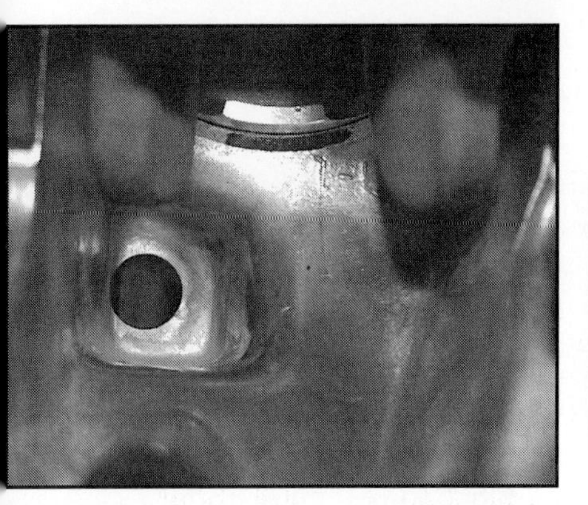

To clear the crank pins on a heavily stroked motor, the block must be milled for clearance. This is an easy procedure that most machine shops can do. A B18C block modified by R&D is shown here.

Here is the clearance notching of a R&D D16 motor.

The block of Adam Saruwatari's NSX is built to take over 1000 hp worth of turbo boost. It features Darton sleeves that also reinforce the deck area to prevent flexing and blown head gaskets.

Even though all but the latest generation of the H22 Prelude motor have a closed stronger deck, this large displacement R&D Racing stroker motor has sleeves to allow a bigger overbore.

reasons for doing this. The first is that the Honda block is made with an advanced high pressure die casting process and the die for the water jacket must be able to be released as the freshly cast block clears the mold. It can only go upward so the block's deck is open. The second reason is that because coolant can freely circulate around the top of the bore, the rings can seal with less cylinder wall distortion. The top of the bore is the hottest and even cooling helps here. Better ring seal means lower emissions. More even temperatures in this area and the combustion chamber also result in lower emissions.

The problem is that when these engines are taken to the limit, at high rpm when naturally aspirated or when boosted beyond 15 psi, the unsupported cylinder walls crack from the side load and combustion pressure-induced stress. The narrow sealing surface of the open deck block can also blow gaskets at high boost levels. This is a shame with the Honda's ultra-strong cranks and bottom end.

The Golden Eagle sleeves as shown in a cutaway of Ed Berganholtze's block have thick walls and a built-in block guard that braces the cylinder walls up against the outside of the block. These suckers won't crack no matter how much boost they are fed.

Block Guards

All is not lost though. To solve this problem, racers have been using several approaches. The cheapest and easiest involves installing a block guard. Nu-Forms and STR make block guards, among others. A block guard is a CNC-machined piece of aluminum that is tapped into the top of the block deck, filling all the voids between the water jacket and the freestanding cylinders. The block guard is supposed to support the cylinders and keep them from moving around or cracking. The block guard has holes drilled in it so cooling water can flow through it to the head.

The block guard is regarded as a good thing by some people and a bad thing by others. I tend to be in the latter category. For one, there are a lot of misconceptions of how to install a block guard due to some misleading articles in the enthusiast press. One article showed a block guard being installed in a completely assembled motor while it is still in the car simply by tapping it in place with a hammer. This sort of installation greatly distorts the upper part of the cylinder bore by up to several thousandths of an inch, causing poor ring seal at best and seizing at worst. One very well known Quick Class competitor had his block and cylinders crack right at the point of where the block guard was installed, probably because of the cylinder distortion and the huge stress riser the block guard created. There were also signs of seizure on both the piston and the cylinder wall in the area of the block guard.

If a block guard was to be used, the best way to install it would be the way JG and other engine builders do it. They hand-fit the block guard to the block for a close fit before any other machining operations are done. The block guard is then TIG welded in place. Skip welding is used to minimize distortion to the rest of the block. After welding they machine the block deck flat, THEN the final bore machining is done. In my opinion, this is the correct way to use a block guard. At the very least, the block's bores need to be machined after the block guard is installed. In my opinion, a block guard should never be installed in an assembled motor sitting in a car with its head off. It should only be installed when building a new motor, and the bore honing can be done after the block guard is installed.

Resleeving

A better way to beef up a Honda's cylinders is to resleeve the block. This involves boring out the thin cylinder wall liner and press fitting in new bore sleeves made of a thicker superior material, like chrome-molybdenum steel. The best sleeves, in my opinion, are those made by AEBS. These sleeves feature a thick deck plate built into the sleeve, which solves both the sleeve cracking and any head gasket sealing issues in one easy operation. The deck plates butt into each other and brace against the outer flanges of the block so they cannot move at all. This gives a superior strong cylinder wall and lots of clamp area for the head gasket, as well as lots of support, preventing the sleeves from walking around. The guys running in the All Motor class use sleeves also for a different reason, to allow a super big overbore for more displacement.

STR also makes excellent thick-wall sleeves that brace against the block wall to prevent shifting under load. The top of the STR sleeve is more open than the AEBS sleeve, which allows more even coolant circulation but does not give as much gasket seating area. This may be better for strong normally aspirated motors, but boosted or nitrous'ed motors would still probably be better off with the thick-decked AEBS sleeves.

Other than redoing the cylinders, the rest of the Honda block is built like a tank and needs very little modification to handle the most extreme power demands reliably. The H-series Honda motors have a closed deck and do not exhibit these problems.

The thick buttressed AEBS sleeves combine beefed-up cylinder walls with the support of a block guard all on one tough unit. These are almost bomb proof.

Here is a thick-wall STR sleeve in a big-bore D16 motor.

This R&D-prepped block shows the smoothness of a proper bore finish for low-tension rings. Much of the smoothness comes from plateau honing. Although an old-school machinist will probably wince when he sees smooth cylinder walls like this, modern rings don't need a rough finish to seat and break in correctly.

Machining Tips

When boring and honing a block, a deck plate should be used. This is a thick metal plate that simulates the distortion caused by bolting on the cylinder head. This helps to ensure that the cylinder bores stay round and the ring seal stays good even once the head is torqued down. Although this is less critical on a beefy import motor than it is on a flexible domestic V-8, and plates for import engines are hard to come by, it should still be done if possible. The cylinder bore can distort as much as 0.0005" on some motors, which is an amount that does make some difference in friction and ring seal. Each piston should be measured and instructions given to the machinist so he can keep the piston-to-wall clearance equal on each individual piston and bore. Open deck blocks like the B- and D-series Hondas don't need a deck plate because the freestanding cylinders do not distort from the bolts. When the deck is modified or closed, then using a deck plate may be beneficial.

Bore Finish—Proper cylinder wall finish is important also. It is critical to use the exact finish prescribed by the maker of the rings for both long life and good power-producing ring seal. As I mentioned earlier, I believe that the stock ring packages that come with most import motors are better than the aftermarket versions. These rings are chrome faced for long life and are usually thin in profile and low

This R&D-prepped big displacement 2500cc B18C has ARP studs holding down the head. The high strength studs allow for greater torque, which creates better head gasket sealing. Although this is a high-compression engine, a high-boost turbo can benefit as the studs can help prevent head lifting and blown head gaskets.

in tension that helps seal at high rpm. For these rings, a smooth cylinder wall finish is required in the 600-grit range. All honing should be done on a Sunnen CK-10 machine, which is the best for accurate work, with diamond grit stones, which control tolerances better. After honing, a few passes need to be made with a plateau hone, which uses cork-bonded grit stones, or a fine grit flex hone. Plateau honing removes the microscopic high peaks in the honed surface. This speeds break in and reduces wear during break in.

Align Boring—Although import motors rarely need it, all blocks should be checked to see if align boring is necessary. Align boring is used to correct the straightness of the main bores if they have changed during the engine's life. Although this is a common operation when building a domestic motor, an import hardly ever needs this done. Nevertheless, it should always be checked when building any motor.

The block should be fully deburred with cartridge rolls and deburring tools to get rid of stress risers, sharp edges and casting flash. All freeze plugs should be staked in place and some engine builders go as far as to set screw them in place. All press-fit oil galley plugs should be replaced with the screw-in type. The really detail minded will polish the block's interior to remove embedded casting sand. This eliminates any possible stress risers, aids oil return and reduces windbag losses.

Some guys don't do any of this and still go fast with lots of power; this is a testimony of the strength and sound engineering behind most import motors!

Before Assembly

Before you begin assembly, it is important to thoroughly clean the block. After machining, the block is heavily contaminated with honing shavings and debris embedded in sticky cutting oil. Sludge builds up in the block's oil passages with use and really piles up in the blind ends of the passages.

All oil passage plugs need to be removed and the passages need to be brushed out with rifle brushes or cleaning brushes, especially designed for engine oil galleys. Summit Racing sells block-cleaning kits for this purpose. A rust-inhibiting degreaser, like Motul Moto wash, should be used on the brushes, in the passages and generously sprayed around to get every bit of crap off. The oil passage plugs must be reinstalled with Loctite™ after cleaning or horrendous internal oil leaks will instantly destroy the new motor! After scrubbing, the block should be flushed with lots of water and blown dry with compressed air, especially the oil passages. Care should be taken to avoid getting water on the freshly machined bores at this point, as they will rust almost instantly.

Once the block is clean, then the bores can be cleaned with carb cleaner and clean white cotton rags until the rag is absolutely white after wiping the bore. After the bore is completely clean, it should be sprayed with a rust-inhibiting oil like Motul Protect to prevent any rust from forming.

Studs

A trick used on many domestic motors is studding. Studding is replacing the bolts that hold the head and main caps in place with high strength threaded studs. A stud has the potential to be stronger than a bolt because it holds the parts in

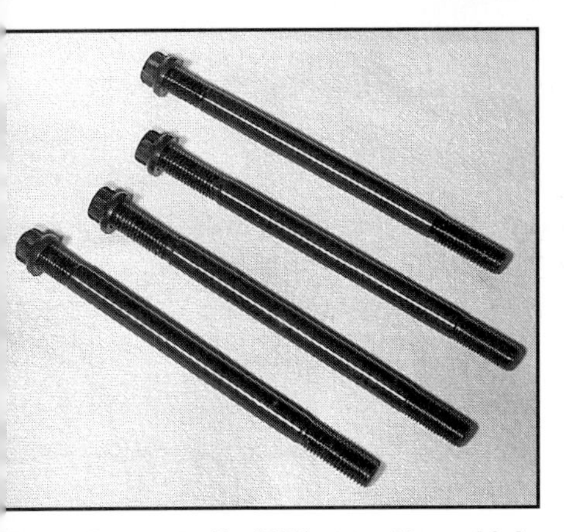

Adam Saruwatari's NSX uses these high-strength ARP bolts to secure the main bearing caps and girdle under the loads of over 1000 hp worth of boost.

compression instead of tension. ARP and SPS make ultra high strength metric studs with various metric threadings for import motors. However, the bolts used for the main caps and cylinder heads of our beloved imports are already pretty high quality and very strong. Failure is rare even with mega-boosted motors and many a 9- and 10-second motor has been held together with stock bolts. Still, if you want to go all-out, you may consider adding studs.

An important point to remember is that many imports use what is called a stretch bolt for these parts. Stretch bolts have a section of the shank with a reduced diameter that stretches when tightened to the correct tension. This ensures the correct tension will be applied. A sure sign that stretch bolts are being used is if the torquing procedure is an angular method of tightening the bolt. This is when the bolt is tightened to a low torque first with a torque wrench, the turned a given amount of degrees after that to the final tightness. Stretch bolts give superior performance but must be carefully inspected and measured for overall length and thread distortion after each use to ensure they are not overstretched or deformed.

THE OILING SYSTEM

Hondas have universally good oiling systems. They feature stout gerotor pumps driven directly off of the crank with no failure-prone drive systems. Little modification is needed even at extreme power outputs. Some folks close up the end gaps of the pumps to reduce internal pump blowby by lapping the housings. Others shim the oil pump relief spring to increase the oil pressure. However, these modifications are really very optional, and most engine builders do nothing in this area.

If you are building a Honda hybrid like an LS motor with a V-Tec head or a C20 block with a V-Tec head, you might consider swapping in the higher capacity oil pump from a V-Tec motor.

Sometimes oil pump rotors can fail. Although it is rare for this to happen, this usually results in a complete failure of the motor. Usually, only really high-revving, all-motor class engines suffers this sort of failure when the motor is revved into the high 9000 rpm range. Unorthodox Racing makes high-strength billet oil pump rotors and pump backplates for Hondas to prevent this failure.

Since most oil systems are engineered so well, the main mods to the oiling system are simple: thoroughly clean the passages and add screw-in galley plugs secured with Loctite to make sure they don't blow off. Chamfering the sharp edges of the galleys on the crank journals and anywhere the oil flows cannot hurt either. Other than that, there is not much you can do.

If you suffer from a blown engine, it is super important to check the oil pump for internal scoring and jammed relief valves and other damage from debris, to clean out all the oil passages fully, and to clean any external oil coolers and lines. Failure to do this has caused many a newly rebuilt

Unorthodox Racing makes an oil pump with billet gears and a steel back plate to prevent oil pump failure, even in a high revving motor.

AEBS makes an innovative electric water pump that does not suffer from cavitation and reduces parasitic power losses. Recent testing by *Turbo* magazine shows that gains of up to 8 wheel hp are possible from the pump.

motor to blow again, right after the fresh rebuild.

Dry Sump Systems

I predict that as the sport of Honda drag racing progresses, dry sump oiling systems will become important to both make more power and to help engines live under extreme conditions. Dry sump systems use an external belt-driven multi-stage oil pump. Usually, two stages of the pump scavenge the engine's bottom end and one stage scavenges the cylinder head. From the pump, the oil travels to a filter, an oil cooler, then finally to a big external tank where it is deaerated. Then the pump's fourth pressure stage sends the back to the engine's main oil galley.

Dry sumps are mandatory for road racing where engines spend a lot of time at high rpm and under high G-loads. The high volume pump also ensures that the engine's bearings and drivetrain are fed ample amounts of oil no matter how severe the conditions are. How do dry sumps produce more power? The powerful scavenging of the three suction stages of the pump suck the oil out of the engine quickly, hence the name dry sump. This cuts down windage losses because the crank doesn't ever splash into oil in the pan nor cut through a heavy oil rainstorm in the crankcase. The efficient suction of the scavenge stages of the pump also produce a vacuum in the crankcase, further reducing aerodynamic losses and helping to improve piston ring seal.

Typically a dry sump system really helps an engine live and produces about 5-10 more hp over a stock oiling system. The complexity and the expense has kept dry sumps out of import drag racing for now, but that may change in the near future as development progresses.

Cooling System

Very little needs to be done to the cooling systems of an import motor. Fluidyne makes aluminum radiators for most of the popular Hondas and sometimes you can run a radiator from another model of car for more cooling capacity. Using an Integra radiator in a Civic is a common trick used by a lot of racers. AEBS has electric water pumps for drag racing Hondas to avoid parasitic power loss and to ensure that a good volume of coolant reaches the far corners of the engine under all speed conditions.

The detail conscious will break the edges on water pump impellers, radius the cooling passages and remove accessible casting flash from the block's water passages. That is about it. Many a fast car has no work done at all in this area.

Running a higher-pressure radiator cap and reducing power-robbing water pump cavitation and overheating area is about all you can do. Some cars may benefit from a lower temperature thermostat although probably not ones used in drag racing. Engines produce the best power when the oil temperature is over 200 degrees and the water over 180 degrees.

HEAD GASKETS

The ubiquitous, often ignored head gasket is an important link in the building of an all out motor. Interestingly enough, the stock Honda head gaskets are perfectly adequate for pretty severe use. Stock head gaskets are usually made of steel shim stock covered with graphite filled composite. Some stock Honda gaskets are even made of tough PFTE coated stainless steel shim stock. These combinations of gasket materials seals well and are reasonably strong, even to large amounts of boost unless the engine is detonated. In that case, the high pressure of detonation pounding can often cause the stock gasket, or any gasket for that matter, to blow out.

Some tuners O-ring the block or cylinder head to help the stock gasket resist blow out. O-ringing is the addition of a machined groove around the circumference of the cylinder bore in either the block or the head where a wire made of either copper or soft annealed stainless steel is inserted by tapping it into place. The fit between the wire and the groove is a tight, interference fit. This wire digs into the head gasket and helps pin it in place to resist blow out. O-ringing should be done before the final honing of the bores as the groove machining process and the inserting of the wire can cause the bore to close up slightly.

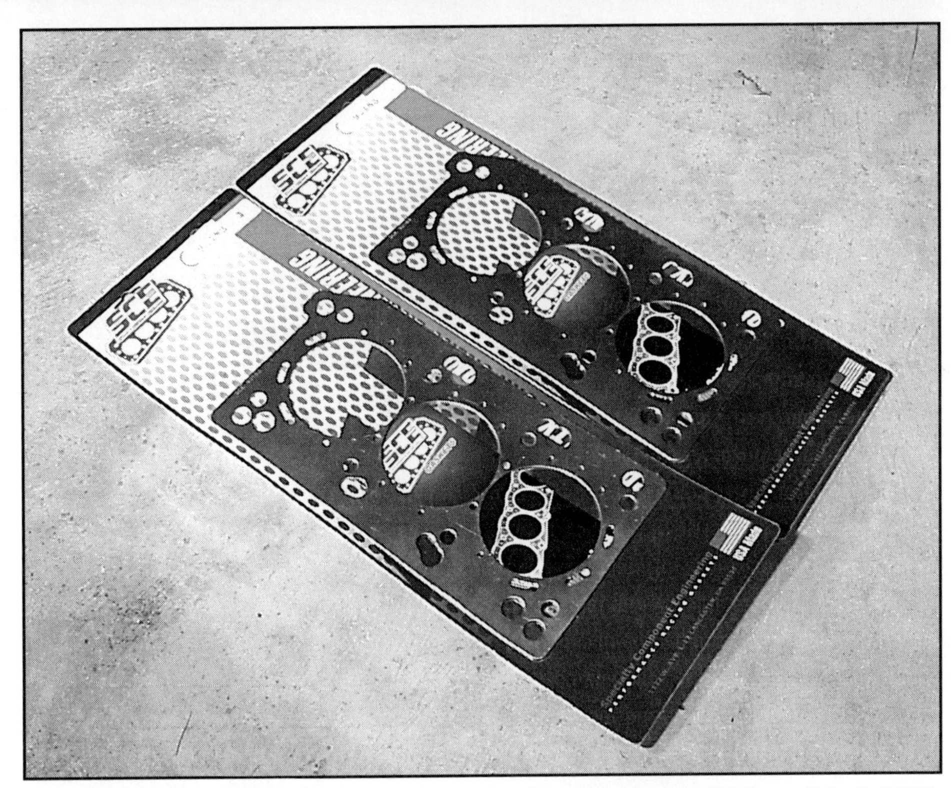

Adam Saruwatari's NSX uses O-ringed copper head gaskets to contain the anticipated high boost pressures.

Aftermarket Gaskets

HKS and Greddy offer head gaskets made of tough, Teflon®-coated steel shim stock. These gaskets are very strong.

For extreme turbo and nitrous use or for extra large sleeved bores where a custom head gasket is needed, soft annealed copper is the material of choice. There are quite a few companies that can laser cut custom head gaskets out of copper. The difficulty with copper is that it is hard to get a good seal with the water jackets. The copper gasket must be re-annealed after cutting to ensure that it is soft and thoroughly coated with Hylomar or a similar semi-non hardening sealant. Even then, copper often leaks coolant slightly, limiting its use to all-out drag racing engines.

Greddy and HKS also sell thick 2mm or sometimes even greater head gaskets for the purpose of lowering the compression ratio for a turbo kit or without spending money on special pistons. Although this is a cheap way to lower compression, it is not the best way to do so. A thicker head gasket is more likely to blow out because there is more surface area around the bore for detonation pressure to impinge upon. A thick head gasket also makes the cylinder head's quench areas ineffective, increasing the cylinders' propensity to detonate more than the suggested reduction of compression the head gaskets would seemingly indicate. A motor that has the compression ratio reduced to 8.5:1 via a thick head gasket may have a detonation threshold of a 9:1 compression motor due to the loss of quench.

5 HAULING HYBRIDS

A Guide to Popular Engine Swaps

A GReddy turbocharged B16A has been grafted into the engine bay of this plain Jane LX Civic.

Humankind has made a habit of one-upping the creator when he can benefit in some material sense. We figured out early on that playing matchmaker to a donkey and a horse produced a creature with the strength and speed of a horse, but the nimble sure-footedness of a donkey. This creature is called a mule if you have not figured that out yet. We as enthusiasts want a similar paradox in our cars—we all want our cars to run like a thoroughbred horse and handle like a donkey, thus the notion of installing a larger motor in a lighter chassis is appealing.

Our parents and grandparents who may have been into hot rodding in the '40s and '50s are very familiar with the concept of building hybrid cars. In their day, that meant dropping a big Caddy motor into the smallest car they could find, be it a chopped and channeled '32 Ford roadster for our grandparents, a big-block Chevy into a Fiat Topolino for our parents or a Prelude H22 motor into a Civic for us! As a side note, almost every Quick Class or All Motor Import Drag Racing competitor has some sort of hybrid Civic racecar, usually to a B18 Integra engine or a less common H22 Prelude engine.

Like many manufacturers who often use a small number of different engines for their entire model line, Honda has produced just a few different four-banger engines. However, there are a large number of different variants of these basic engine families. Many American Honda engines have a brother with the same engine designation but with a higher-output in the Japanese domestic or European market. This gives Honda owners quite a smorgasbord of engines to choose from. Some motors, like the vaulted B18 C-5 Integra Type R powerplant, are scarcer and more desirable than others, leading to an unfortunate amount of Honda theft. Because of this widely growing problem, be sure to buy your motor only through legitimate sources as the police are on to Honda theft and many an unsuspecting Honda hybrid owner has been arrested and has had his car impounded for having a stolen motor.

Although hybrids are popular across the Honda line, the most popular ones involve removing the lesser D and EW motors out of Honda Civics of the 3rd generation (84-87), the 4th generation (88-91), 5th Generation (92-95) and finally the 6th generation (96-00) to install something more potent.

Removing the basic duty EW and D-series motors to swap a higher performance Honda motor into the lightweight Civic chassis is an awesome way to get the usually oxymoronic combination of power, driveabilty and reliability out of a Civic. Some of these combos can produce performance of cars costing many times more while retaining daily driver reliability. It is possible to make a daily driver, stock car reliable, low 13-second quarter mile car, so tame, non-finicky and tractable that

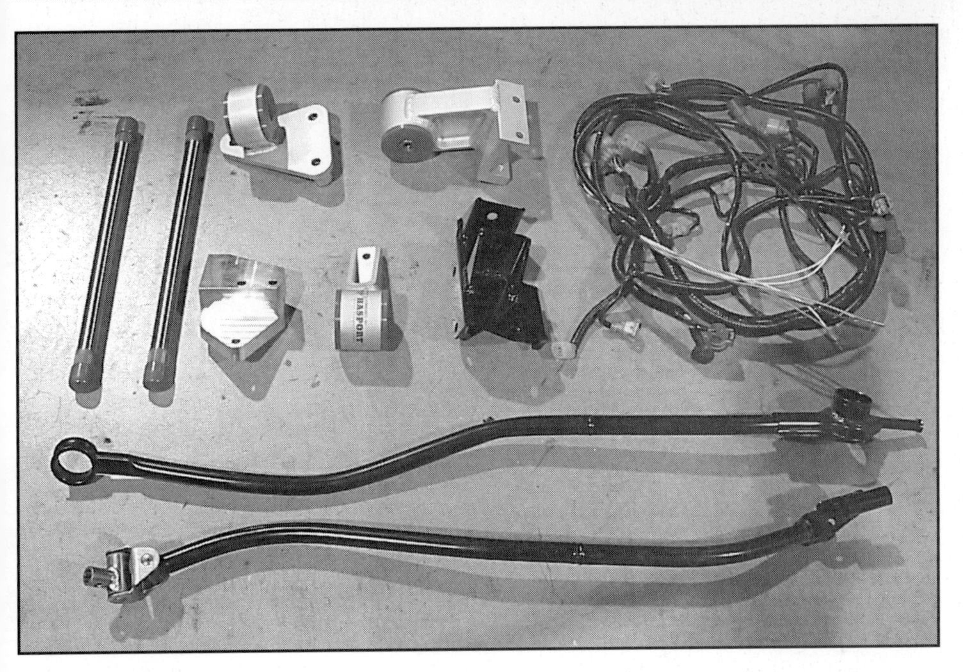

Hasport makes perhaps the most comprehensive engine swap kit on the market with everything included to make building a factory-like-running hybrid a weekend project. A Hasport kit with some of the information in this chapter of the book makes building your own hybrid a relatively easy project.

you could loan it to your grandmother for a week and she would not complain. It could also pass your local smog test for hassle-free registration. The cool thing about a properly built hybrid is that you can have it all, performance with no hassles.

If you desire even more streetable horsepower, there are other tricks that you can do to get even more horsepower out of your engine and still keep streetabilty. There are many parts in the Honda factory parts bin that can be combined to make hot combinations, combinations that Honda never dreamed of doing but are possible with a little creativity. Our next chapter will talk about these streetable engine mods.

I would like to thank the editors of the Honda Hybrid web page, http://hybrid.honda-perf.org/ for assembling an awesome collection of facts on how to do Civic swaps. Navigating to this page is essential for any Honda enthusiasts reading this book. Brian Gillespie of Hasport was also very helpful in helping with the research needed for this difficult section.

THIRD GENERATION CIVIC SWAPS, 1984 to 1987 CIVIC/CRX

The third-generation Civic was the first Civic to be taken seriously by the performance enthusiast. This model got away from the goofy styling of the previous models, and for its time, offered world performance and handling in a lightweight chassis. Even today, it is still a common sight to see a third gen going toe to toe with newer vehicles at your local autocross or SCCA IT race. During this time the CRX was also born, a zippy two-seat coupe based on the Civic.

Due to their light weight, these Civics still performed well even with their wimpy 62–108 hp EW13 and EW15 motors. The 84–87 is the lightest Si.

Below are listed some of the common and worthwhile swaps for this generation that are easy and can be done by a person with reasonable mechanical knowledge.

1986-1987 Integra 1.6L, Japanese Domestic Market ZC

The D16A1 motor from the first generation 86–87 Integra is an easy swap, so easy that it pretty much bolts right in. The Integra motor packs 112–118 hp and a 7500-rpm redline. The JDM (Japanese Domestic Market) version of this motor is called a ZC, which is the same except for some minor differences, mostly slightly hotter camshafts, which give it a bit more power (up to 130 hp total). The ZC is from the Japanese and European market 85–87 CRX. These motors are relatively inexpensive from a wrecking yard or a used Japanese engine broker. The fact that these motors were designed for this car makes the swap much more straightforward.

It should be possible to convert a CRX 1.5 EW carbureted model to the fuel-injected, twin-cam motor but the fuel tank, fuel pump, fuel lines, wiring harness and ECU will be needed for the conversion. Thus, it is easier to start with a fuel-injected model like the Civic or CRX Si. The Si model has better suspension and brakes to handle the extra power and weight of the twin-cam motor stock.

Parts Needed—Here are the parts needed from an Integra to make the Integra or ZC twin cam swap a bolt-in affair. To avoid the nickel-and-dime situation, it might be better to buy a theft-recovery Integra with an intact drivetrain or a JDM front clip. With these scenarios, you can also salvage some trick suspension and brake parts to make the conversion more sano and factory-like. The parts are:

• 1.6L engine, incl. manifolds
• 5-speed transaxle, including axle shaft extension and bearing
• Left and right axles, new axle shaft nuts
• Left and right knuckles and hubs
• Rear engine mount bracket
• Transmission mount
• Wheel hubs
• Air cleaner-to-throttle body hose, w/clamps
• Upper radiator hose
• Throttle cable

These are some of the extra parts that would be nice to salvage from the Integra if you do have access to the whole car or a front clip. Adding the Integra drivetrain will add about 45 lbs. to the front of your Civic, so if possible, it is nice to have the stronger front crossmember and stiffer torsion bars to make up for the increase in weight. Although it is not needed, it is possible to adapt the Integra's power steering with the parts listed below:

• 23mm aftermarket torsion bars are recommended if using a Civic crossmember, to handle extra weight.
• The Integra front crossmember, w/torsion bars and antisway bars are required for power steering. The Integra crossmember is also stiffer and stronger which should help handling.
• Integra steering rack is needed if you wanted power steering, which is not really necessary.

Engine Mount—Finally the Civic front engine mount must be modified to get everything to bolt together. The necessary modification to the engine mount is easy to do. A simple self-explanatory notch must be cut out of it to clear the bigger cylinder head of the Integra motor. Depending on the model and the sagged condition of the motor mounts, the hood ribbing may need to be modified for extra clearance between the hood and the engine's timing-belt cover.

Hasport motor mounts are configured for easy bolt-on operation. They are made of high quality CNC machined and fabricated aluminum alloy with polyurethane mounts. These mounts are very helpful if you do not know how to fabricate.

Electronics—Since the Integra and the ZC engines are fuel-injected, an Integra ECU is needed if you don't want to use or don't have the Civic Si ECU. The engine will still run on the Si ECU. It is better to run the Integra ECU if it is available to enjoy the higher redline that the twin-cam engine is capable of. The Integra redline is 7500 vs. the Si's lower 7000 rpm redline. For the Integra ECU just plug it in and you have two choices on the wiring. If you choose to use the stock Civic Si harness you will need to route the Idle solenoid wiring to the IACV (EACV). The Idle solenoid is in the emissions control box on the driver's side of the engine bay. You will need to tap into the Black/Yellow wire for EACV power and then route the Green/White wire to the EACV for ECU control signal. If you have the EACV plug and some wire from the Integra harness, Black/Yellow goes to Black/Yellow and Green/White goes to the other.

The other choice is to use the 86–87 Integra wiring harness. To use this harness, you must switch the oil pressure sending wire, water temp sending wire and the reverse light wires to match the Civic locations on the passenger side of the harness on the rectangular plug. On the driver's side, you move the above-mentioned Green/White wire to the only open position on the six-pin connector that hooks to the engine harness. This will give you EACV control.

If you have a ZC engine, the wiring solutions will depend on whether the engine originally came in a JDM Civic or Integra. The 85–87 Civic ZC does not have EACV, so just use the stock harness. The Integra ZC is like the 86–87 US Integra.

The Integra injector (control) resistor box is not needed, the one from the Civic works just fine.

A/C—If your car already has A/C and you want to retain it, it is best to use the Civic compressor so you don't have to change to custom A/C hoses. It is also a better design.

Another cool thing about the ZC is that Jackson Racing makes a supercharger kit especially for the ZC Civic Hybrid since it is such a common, inexpensive and easy swap.

B-SERIES TRANSMISSION GUIDE

When swapping out your old EW or D motor for a hotter B motor, you also need to swap out the transmission for a B-series tranny. Here is a guide to the many confusing varieties of B-series transmissions. This tranny chart will help you understand the transmission designations mentioned in the main charts on swapping that follow. Here are a few pointers:

You don't want to install a wide ratio US market GSR type tranny with the 1.6-liter B16A because the wide ratios will really bog down the higer revving, less torquey motor. The closer ratio boxes are probably better for most applications except for turbo-powered drag cars that can use the wide gear spacing.

All Type R transmissions, either USDM or JDM have LSDs (limited slip differentials). These are worth the extra price you pay. All LSD trannys are ink-stamped with LSD on their case. Another way to spot an LSD tranny is to look down the axle hole in the tranny's case. If you can see all the way through, it's an LSD tranny. If you see a bar inside blocking your view, it's a standard tranny. Unfortunately, many unscrupulous people often rip off a tranny's LSD or make a fake LSD ink stamp to sell a used tranny for less. Always make sure you can see through the axle hole on an LSD stamped tranny.

The JDM 98-01 Integra Type R transmission has a 4.75 final drive ratio, which is pretty sweet for acceleration, drag racing and most road racing, but not so good for freeway cruising. Hasport has a kit to convert a cheap JDM YS1, Y1, S1, J1 and YS1 cable equipped tranny to the USDM hydraulic clutch. This is a common, cheap and plentiful tranny in import junkyards, because before the Hasport kit came about, there was no market for them here. Any of these trannys in the transmission chart will bolt to any B-series motor:

B-SERIES TRANSMISSION GUIDE

Tranny Code	S80	Y21/S21/ S4C	Y21/Y80/ S80/S4C	S80/Y80	S80	YS1	Y1	S1/J1	S1/YS1
Chassis Grade	JDM 98-01 ITR	US 92-01 Delsol+Si Coupe	US 97-01 ITR JDM 95-97 ITR JDM 92-00 SiR/CTR	US 94-01 GSR JDM 93.5-01 SiR/SiG	US94+ LS/GS/ Spec.Ed.	JDM 92-93 Integra RSi/Xsi	JDM 89-91 Civic/CRX SiR	JDM 90-91 Integra Rsi/Xsi	US 90-93 Integra RS/LS/GS Spec.Ed
Tranny Type	Hydro	Hydro	Hydro	Hydro	Hydro	Cable	Cable	Cable	Cable
1st	3.230	3.230	3.230	3.230	3.230	3.307	3.166	3.230	3.230
2nd	2.105	2.105	2.105	1.900	1.900	2.105	2.052	2.105	1.900
3rd	1.458	1.458	1.458	1.360	1.269	1.459	1.416	1.458	1.269
4th	1.034	1.107	1.107	1.034	0.966	1.107	1.103	1.107	0.966
5th	0.787	0.848	0.848	0.787	0.714	0.875	0.870	0.848	0.742
Reverse	3.000	3.000	3.000	3.000	3.000	3.000	3.000	3.000	3.000
Final Drive	4.785	4.400	4.400	4.400	4.266	4.400	4.266	4.400	4.266

Special Comments

JDM 987-01 ITR: S80—This tranny has a low 4.785 final drive ratio.

US 92-01 Delsol+Si Coupe: y21/S21/S4C—No LSD (limited slip diff)

US 97-01 ITR, JDM 95-97 ITR, JDM 92-00 SiR/CTR: Y21/Y80/S80/S4C—All "R" trans are LSD equipped. SiR models have optional LSD. Look for LSD stamp on case.

US 94-01 GSR, JDM 93.5-01, SiR/SiG: S80/Y80—US GSR are non-LSD. JDM SiR/SiG has optional LSD. Look for LSD on stamp case.

US 94+ LS/GS/Spec. Ed: S80—Non-LSD. Gear rations make for excellent gas mileage.

JDM 92-93 Integra Rsi/Xsi: YS1—Same gearing as 92-01 hydro SiR/ITR/CTR but in cable form. Optional LSD.

JDM 89-91 Civic/CRX SiR: Y1—Y1's have optional LSD. Check for LSD stamp on case.

JDM90-91 Integra RSi/XSi: Si/Ji —Non-LSD.

US 90-93 Integra RS/LS/GS Spec. Ed SI/YSi—Non-LSD.

The Big B motors: B16A: 1993-1999 Del Sol VTEC, 1989-2000 JDM Civic Si and Type R, B18B-B18A: 1990-2001 Integra LS and B18C, 1994-1996 Integra GS-R

For the really hardcore that want the maximum power in the smallest, lightest chassis, any engine from the B-series family can be easily bolted into the third generation with a Hasport kit. The kits are not absolutely necessary but since the B motors and transmissions are so different in size and mount construction from the old D motor, one must have an extensive knowledge of both automobile systems and fabrication if attempting this swap without one. Even if you are good at these things, the price of the Hasport kit probably makes it more cost effective to use it rather than reinvent the wheel yourself. Hasport's kit is very complete with all of the mounts, axles, linkages wiring harnesses and even instructions to make this involved swap a weekend affair. If you read our chapter on streetable engine mods, it is clear that a very slick machine can easily be built with this kit and a warmed-over B Series motor!

As there is a very mature aftermarket for the B motor, it is possible to make a very high-powered, lightweight car that can propel a little Civic into the high 12's and low 13's and still be a reliable daily driver. These swaps make for the ultimate sleepers.

Swap Info 3rd Generation 1984-1987 CRX/Civic Si with 1986-1987 Integra or JDM Civic ZC engine

Transmission: 86-89 Integra or JDM Civic (86-89 brown valve cover) style ZC
Mounts: Use combination of stock Integra and Civic/CRX mounts
ECU: 86-87 Integra PG7
Axles: 86-89 Integra axles and 86-89 Integra knuckles and hub assemblies.
Shift linkage: 86-89 Integra
Electrical differences: Add EACV
Upper radiator hose: 86-89 Integra, trim to fit
Lower radiator hose: 86-89 Integra, trim to fit
Throttle cable: 86-89 Integra w/ manual tranny
Clutch cable: 86-89 Integra
Air conditioning: Use stock bracket and compressor.
Special instructions: None
Special considerations: Easiest, 86-87 Integra 112hp, ZC 130hp

Swap Info 3rd Generation 1984-1987 CRX/Civic Si with 1988-1989 Integra or JDM Integra ZC engine

Transmission: 86-89 Integra or JDM Integra (86-89 black valve cover) style ZC
Mounts: Use combination of stock Integra and Civic/CRX mounts
ECU: Recommend Zdyne ECU for use with 85-87 Si and this engine. May use 88-89 Integra PG7 with proper electrical modifications.
Axles: 86-89 Integra axles and 86-89 Integra knuckles and hub assemblies.
Shift linkage: 86-89 Integra
Electrical differences: For use with ZDYNE ECU add EACV, Cylinder sensor, ECU memory and Igniter signal. For use with PG7 also add VSS and ELD. Custom Hasport conversion harness available for use with ZDYNE modified ECU.
Upper radiator hose: 86-89 Integra, trim to fit
Lower radiator hose: 86-89 Integra, trim to fit
Throttle cable: 86-89 Integra w/ manual tranny
Clutch cable: 86-89 Integra.
Air conditioning: Use stock bracket and compressor.
Special instructions: Need to have good understanding of electrical system.
Special considerations: Second easiest, 88-89 Integra 118hp, ZC 130hp

Swap Info 3rd Generation
1984-1987 CRX/Civic Si with 1st generation 1989-1991 JDM Civic/Integra B16A engine

Transmission: Japanese or American market S1, J1, A1, Y1 or YS1

Mounts: Hasport Mount kit M84-B16-10

ECU: Recommend Zdyne one wire conversion ECU for 85-87 Si. May use PR3 or PW0 with proper electrical modifications.

Axles: Hasport custom axle shafts used with 86-89 Integra inner and outer joints and 86-89 Integra knuckles and hub assemblies.

Shift linkage: Hasport custom linkage, lengthened stock linkage, or lengthened 86-89 Integra linkage.

Electrical differences: For use with ZDYNE ECU add EACV, Cylinder sensor, ECU memory Igniter signal and VTEC solenoid. For use with PR3/PW0 ECU's, add VSS, VTEC oil pressure, knock, and secondary O_2 sensors. Custom Hasport conversion wire harnesses are available for use with ZDYNE one wire conversion ECU.

Upper radiator hose: 92-93 Integra GSR, trim to fit.

Lower radiator hose: 90-91 Integra, trim to fit.

Throttle cable: 90-91 Integra

Air conditioning: No

Special instructions: Need to put a large dent in frame rail for alternator clearance.

Special considerations: All parts will run about $2500 to $3000.

Swap Info 3rd Generation
1984-1987 CRX/Civic Si with 1990-1993 Integra LS B18A, 1994-1996 Integra LS B18B, 1994-1996 Integra GS-R B18C engine

Transmission: Japanese or American market S1, J1, A1, Y1 or YS1

Mounts: Hasport Mount kit M84-B16-10

ECU: Recommend Zdyne ECU for use with 85-87 Si and this engine. May use PR4 with proper electrical modifications.

Axles: Hasport custom axle shafts used with 86-89 Integra inner and outer joints and 86-89 Integra knuckles and hub assemblies.

Shift linkage: Hasport custom linkage, lengthened stock linkage, or lengthened 86-89 Integra linkage.

Electrical differences: For use with ZDYNE ECU add EACV, Cylinder sensor, ECU memory and Igniter signal. For use with PG7 also add VSS and ELD. Custom Hasport conversion harness available for use with ZDYNE modified ECU.

Upper radiator hose: 90-91 Integra, trim to fit.

Lower radiator hose: 90-91 Integra, trim to fit.

Throttle cable: 90-91 Integra

Clutch cable: 88-91 Civic

Air conditioning: No

Special instructions: Need to put a large dent in frame rail for alternator clearance.

Special considerations: Smog legal

4TH GENERATION CIVIC CRX 1988–1991 ENGINE SWAPS

This Civic was the first to get the sophisticated multi-link suspension. This suspension makes a well-built Civic a world-class cornering machine with the right mods. This is also the last late and great CRX. These are the first Civics capable of very serious performance. The styling of these model years still looks crisp and modern, even by today's standards. In my opinion, this generation is the first of the truly balanced, high performance versions of the Civic and CRX. As the highest powered variants of this model series still only packed a paltry 112 hp, swaps can bring these cars up to speed and handling that their chassis are capable of.

1986-1987 Integra 1.6L, Japanese Domestic Market ZC

The ZC, the standard motor in the 1985-1987 JDM Civic and CRX Si, is a common motor that is plentiful and cheap from Japanese-market used engine importers these days. The D16A1 motor found in early 1986-1987 Integras is also a common motor that swaps easily into these cars. The D16A1 is nearly identical with a very simular swap. Like in the 1984-1987 Civic, the ZC-D16A1 is an easy swap, nearly bolting right in. This is because this chassis was designed to use this motor in other markets. The ZC usually handily comes from the junkyard with 4G Civic mounting brackets on it. The ZC has slightly hotter cams, which gives it a peak power of around 128 hp. Few mechanical mods are needed to install a ZC or a D16A1 in this generation Civic. The motor looks really clean in the engine bay and your A/C compressor should bolt right. The DX throttle cable is long, but will still work. Or replace the throttle cable with one from an Si.

Gearset Options

Gear	1st	2nd	3rd	4th	5th	Final
US Civic Si	1.894	3.250	1.259	1.107	0.771	4.25
JDM ZC	1.944	3.250	1.346	1.107	0.878	3.88
Hybrid Trans	1.944	3.250	1.346	1.107	0.878	4.25

As with the 3rd generation Civic, Jackson Racing makes a supercharger kit especially for the ZC Civic hybrid.

Another good thing about the ZC and D16A1 is that they can bolt straight to the transmission that came in the car making this a very easy swap. Both the Si and DX transmission bolts directly to these motors for a simple, headache-free installation. The DX transmission will sacrifice a little acceleration for slightly better fuel mileage and lower freeway rpms. If you are also going to swap transmissions at the same time, the Si transmission has the best gear ratios and a lower final drive ratio. The JDM ZC transmission has a close ratio 1-5th gear, but a high 3.88 final drive, making the best transmission still the US Si for all around driving fun. For the ultimate acceleration take the Si final drive and combine it with the ZC gearset (see chart above).

The ZC transmission requires the use of ZC or 90-91 Integra axles, and a ZC or 86-89 Integra intermediate shaft. This is a significant advantage especially with higher power modified motors as Integra axles are equal length and will go a long way to reduce torque steer.

For the electronics side of things, you can use an Si ECU (7000 rpm fuel cutoff), 88-89 Integra ECU (7500 rpm fuel cutoff) or the JDM ZC ecu (7500 rpm), unfortunately the DX ECU will not work. For a modified ZC with header, exhaust and intake, the extra power and improved top end breathing makes the Si ecu's redline seem too low. With bolt-on mods, the Integra, ZC ECU's redline of 7500 rpm is just right. With a repro-grammed ECU keep the rev limiter below 8000 rpm or valve float with the stock springs becomes a problem. On the DX, the Integra, Si and ZC ecu will plug into the DX wiring harness, but a few wires will have to be changed.

4G DX to ZC Wiring Conversion (Courtesy http://hybrid.honda-perf.org/)

In order to convert your batch-fire injection DX to multi-point injection, you will need four injector plugs, an injector resistor box, extra wire, a wire stripper/cutter and some heat-shrink tubing. When doing this conversion, it is highly recommend that you solder all connections and cover them with heat shrink tubing. It is also extremely useful to have an extra multi-point injection engine harness laying around to take plugs and ecu connector pins from (i.e. injector plugs, injector resistor box plug, and distributor plug). Listed below is a straightforward and easy-to-follow set of instructions to help you complete your DX to ZC swap:

Converting from a DX to a ZC from the Main ECU Connector:

- Pins B10 and B12 are empty. Unused pins can be taken from B2 or B11. Some models have a wire located at B12.
- Cut orange and white wires off at C1 and C2 and connect them to wires added at B10 and B12. Orange-B10. White-B12. Leave enough wire for next step.
- Run wires from C1 and C2 (direct ECU connection) into the engine compartment. Label these wires for later use.
- Cut wires at A3 and A7 leaving plenty of wire. Run these wires into the engine compartment and label them for later use.

Inside the Engine Compartment

- Transfer DX wiring harness over to ZC engine.
- TPS and EACV plugs are too short and must be extended.
- Be sure to switch green/white and yellow/white wires on TPS injectors and injector resistor box.
- Connect the yellow/black wires from the two DX injector harnesses and run it to the yellow/black wire on the injector resistor box.
- Connect the yellow wire from the DX injector to the #1 injector (brown wire).
- Connect the red wire from the DX injector and run it to the #3 injector (blue wire).
- Connect the A3 wire to the #2 injector (red wire).
- Connect the A7 wire to the #4 injector (yellow wire).
- Connect the 4 red/black wires from injector resistor box to each injector.
- Cylinder position sensor.
- Connect C1 to blue/green wire on cylinder position sensor plug.

B16A1 Wiring Harness Guide

Engine Sensor_	Color	ECU Pin Number
VTEC Solenoid	Yellow/Green	A4
VTEC Oil Pressure Sensor	Blue (pass. side harness)	D6
VTEC Ground	Black	chassis ground
Knock Sensor	Red/Blue	D3
Evap Purge Power	Yellow/Black	12V power (hot in ON or START)
Evap Purge Solenoid	Red	A20

For the B18C1, check on the following 7 wires to see if everything is there and if they are hooked up right.

CU Side Pin/Connection		
VTEC Solenoid	Yellow/Green	A4
VTEC Oil Press. Sensor	Blue	D6
VTEC Ground	Black	Chassis ground
Knock Sensor	Red/Blue	D3
EVAP Purge Power	Yellow/Black	12V Power
EVAP Purge Solenoid	Red	A20
Secondary Intake Solenoid	Pink/Blue	A17

- Connect C2 to blue/yellow wire on cylinder position sensor plug.

That's all you need to get the motor going, the motor should start right up if you did everything correctly.

D16Z6 SOHC VTEC

This is a less popular swap, but it is still a good one. The later model 125 hp SOHC D16Z6 VTEC motor from a 5G Civic Si/EX will bolt to the 4G transmission allowing it to drop into the car with no cutting or welding. Use the ECU and wiring harness from the SOHC VTEC for best results. The 96-and-later SOHC VTEC motor has a better combustion with quench zones and is a little better choice for hopping up.

Bolt-In Kits for B Engine Power

Hasport makes a very high quality bolt-in kit to install a potent B-series motor in the 4G chassis. The kit is very complete with mounts, linkages, wiring harnesses and comes with everything you need to make the completion of the project possible in a weekend. A/C brackets are optional. Perhaps bolting in a B Series motor with a kit is probably one of the most satisfactory swaps making the biggest difference for the least amount of work. A B-series engine, gearbox and ECU are needed with the kit to complete the installation. Possible engine donor cars are:

- B16A: 1993-1999 Del Sol VTEC, 1989-2000 JDM Civic Si and Type R
- B18B-B18A: 1990-2001 Integra LS
- B18C: 1994-2001 Integra GS-R, B18C5 1999-2001 Integra Type R.

As the hop-up potential of the B-series motor is well-known, the performance potential of this swap is simply awesome. For the 1997-2000 Civic Si and the 1997–2001, Integra GS-R as well as the1999–2001 Integra Type R engines, the swap is more difficult and involved as these motors are OBDII equipped. They will easily bolt in but they require the ECU and wiring harness from the same year car to work correctly. The swap is very possible, however. Simply check out the chart for the earlier years of the same motor and just remember that the matching engine harness and ECU must be used with the OBDII motors. HPC and Place Racing also produce high quality bolt in kits for B-series engine conversions.

Swap Info 4th Generation
1988-1991 CIVIC/CRX
1988-1991 CRX/Civic Si with 1988-1991 Civic Style ZC Engine

Transmission: All L3 transmissions are compatible with this engine. They include 88-91 Civic and CRX DX, LX, EX, Si and Japanese ZC.

Mounts: Stock

ECU: PM6, PM7, or 88-89 Integra PG7

Axles: Stock axles with American L3 trannies. For Japanese L3 ZC trannys use ZC intermediate shaft and 2 right-hand 88-91 Civic/CRX axles with 90-93 Integra inner joints or 90-93 Integra axles.

Shift linkage: Stock

Electrical differences: For EX, Si and HF the distributor wiring needs to be slightly modified, for DX, LX, and Standard models the injector and distributor wiring needs to be modified to the equivalent of the Si wiring. Hasport custom wire harness are available.

Upper radiator hose: 86-89 Integra

Lower radiator hose: Stock

Throttle cable: 88-91 Si

Clutch cable: Stock

Air conditioning: Stock

Special instructions: None

Special considerations: None

All information in these tables is courtesy of Brian Gillespie of Hasport

Swap Info 4th Generation
1988-1991 CRX/Civic Si with 1st generation JDM 1989-1991 Civic/Integra B16A engine

Transmission: Japanese or American market S1, J1, A1, Y1 or YS1

Mounts: Hasport M88-B16-10

ECU: PR3, PW0, or ZDYNE One Wire Conversion

Axles: The proper axles can be constructed using 86-89 Integra axles with 90-93 Integra inner joints. If the 90-93 Integra intermediate shaft is used some people opt to use 90-93 Integra axles.

Shift linkage: Hasport custom linkage or shortened 90-93 Integra

Electrical differences: DX, LX, Standard models need to be modified to Si first. To use PR3 or PW0 four wires need to be added, VTEC, VTEC oil pressure, knock sensor and second O_2. To use ZDYNE add only VTEC wire.

Upper radiator hose: 92 GSR, trim to fit

Lower radiator hose: 90 Integra, trim to fit

Throttle cable: 90 Integra

Clutch cable: Stock

Air conditioning: Use Hasport AC bracket with stock AC compressor.

Special instructions: Make a dent on the left-hand frame rail for alternator pulley clearance.

Special considerations: Best bang for the buck, difficult to make smog legal in CA

Note: The Zdyne ECU mentioned in the swap info above and on the next page is a special stock ECU modified to be user programmable like a stand-alone aftermarket unit. We will talk more about the Zdyne in later chapters.

Swap Info 4th Generation
1988-1991 CRX/Civic Si with 1990-1993 Integra LS B18A, 1994-1996 Integra LS B18B engine, 1997-2000 Integra LS B18B Engine

Transmission: Japanese or American market S1, J1, A1, Y1 or YS1

Mounts: Hasport M88-B16-10

ECU: PR4

Axles: The proper axles can be constructed using 86-89 Integra axles with 90-93 Integra inner joints. If the 90-93 Integra intermediate shaft is used some people opt to use 90-93 Integra axles. I would suggest changing the outer joints to that of the 86-89 Integra.

Shift linkage: Hasport custom linkage or shortened 90-93 Integra

Electrical differences: For Si, EX, and HF models, modify the engine harness to fit, for DX, LX, and Standard models the injector and distributor wiring needs to be modified to the equivalent of the Si wiring. Hasport custom wire harnesses are available. (Back date electronics to 90-91 style B18A)

Upper radiator hose: 90 Integra, trim to fit

Lower radiator hose: 90 Integra, trim to fit

Throttle cable: 90 Integra

Clutch cable: Stock

Air conditioning: Use Hasport AC bracket with stock AC compressor.

Special instructions: Make a dent on the left-hand frame rail for alternator pulley clearance. (For B18B, must change left side engine bracket to one from 90-93 Integra to use Hasport mount kits.)

Special considerations: Nice motor, and excellent base for turbo car. Smog legal in CA.

Swap Info 4th Generation
1988-1991 CRX/Civic Si with 1994-1995 Integra GS-R B18C, 1996-2000 Integra GS-R B18C including B18C5 Type R Engine

Transmission: Japanese or American market S1, J1, A1, Y1 or YS1

Mounts: Hasport M88-B16-10

ECU: ZDYNE Two Wire Conversion or ECU that came with the engine with proper modification to wiring.

Axles: The B18C axles can be used in most cases. I would suggest changing the outer joints to that of the 86-89 Integra. The proper axles can be constructed using 86-89 Integra axles with 94 and up Integra inner joints.

Shift linkage: Hasport custom linkage or shortened 90-93 Integra or shortened 94 up Integra.

Electrical differences: DX, LX, Standard models need to be modified to Si first. To use ZDYNE add VTEC and air intake bypass wires. To use engines stock ECU many changes need to be made including changing plugs and the use of an ECU adapter harness.

Upper radiator hose: 94 GSR, trim to fit

Lower radiator hose: 90 Integra, trim to fit

Throttle cable: 94 GSR (or Type R for Type R engine)

Clutch cable: Stock

Air conditioning: Use Hasport AC bracket with stock AC compressor.

Special instructions: Make a dent on left-hand frame rail for alternator pulley clearance. Must change left side engine bracket to one from 90-93 Integra to use Hasport mount kits.

Special considerations: Electrical is difficult, but and excellent choice of engines. American engines installed properly are smog legal in CA.

FIFTH GENERATION, 1992–1995 CIVIC AND DEL SOL ENGINE SWAPS

The EG fifth-generation Civic was made from 92–95 and is considered by many to be the best-looking and having the best potential for modification of all the Civics. This is due to the fact that the motor mounts between the EG and the third generation Integra are identical and that these cars were built before the pesky OBDII emissions systems were required. The wiring harness is also nearly identical so an engine swap is pretty close to plug and play.

The EG is capable of handling the awesome performance potential that the B series can produce as well as reaping the benefits from the tons of racing parts made available for these motors. A streetable B18C motor in a pretty mild state of tune can push an EG through the quarter in the high 12s.

The mighty 1994–2001 third-generation Integra B series motors and transmissions almost bolt right in with little fanfare. In fact, the 1996 Del Sol VTEC B16A2 bolts right into the stock transmission. Because of this, the EG is the cleanest, most factory-like swap you can make. To make the swap, you will need the engine, transmission, axles, shift linkage, ECU and wiring harness from a donor Integra. B motor powered swap donor candidates are:

- B16A: 1993–1997 Del Sol Vtec, 1989-2000 JDM and USDM Civic Si and Type R
- B18B-B18A: 1990–2001 Integra LS
- B18C: 1994–2001 Integra GS-R, B18C5 1999-2001 Integra Type R

For the 1997–2000 Civic Si and the 1997-2001, Integra GS-R as well as the1999–2001 Integra Type R engines, the swap is more difficult and involved as these motors are OBDII equipped. They will easily bolt in but they require the ECU and wiring harness from the same year car to work correctly. This makes the swap more complicated.

The EG has Honda's superior multilink suspension at all four corners which makes it have the potential for world class handling with the selection of the right aftermarket parts. With a strong B series motor and proper suspension and brake mods, the little EG can give supercars fits. With its good looks, light weight and ease of modifying, it is the EG that will become the classic Civic, a favorite for years to come.

General Wiring Tips

The easy part about this swap is that the factory wiring harness can be used. Almost all the wiring plugs in. For GS-R and Del Sol VTEC motors, the Civic wiring harness is missing the knock sensor wires, so this wire will have to be added. Otherwise, there should be few concerns with wiring.

It is critical for proper VTEC operation that the knock sensor be connected, as the ECU will not turn VTEC on unless it sees a signal from the knock sensor. For the B16A1 engine, there are six wires in the harness that need to be verified. See the wiring harness chart nearby.

1993–2001 Prelude H22A1/H23A1 Motor Swaps

For the truly hardcore individuals, Hasport makes a kit so the mighty big-block Honda engine, the H22A1/H23A1 Prelude motor, can bolt right into the tiny EG chassis. Packing 190 hp and a buttload of torque in stock condition, the H motors have the potential to boot a Civic through the quarter mile in the high 12's in nearly stock condition.

Since the H motor adds a bit of understeer promoting weight to the front of the car, the B motored cars are considered to be a bit more graceful around a road course. However with some of the new coil over suspension systems that can make use of a wide variety of different rate 2.5 inch racing springs, this can be largely compensated for.

Hasport can provide all of what it takes to put one of these monsters under the hood of your EG in a most sano, "no fabrication needed" way over the course of a weekend. Place Racing also has a nice kit to install the H22/23 as well as the F22 Accord engine in the EG.

Hasport even has the proper brackets to hook up your A/C. Since Hasport is located in Arizona, it figures that they would develop this part!

Swap Info Fifth Generation 1992-1995 Civic and Civic Del Sol with 1st Generation 1989-1991 JDM B16A, 1992-1993 Integra GS-R B17A, and 1990-1993 Integra LS B18A Engine

Transmission: These engines come with Japanese or American market cable clutch transmissions including S1, J1, A1, Y1 or YS1. To use them you will need the Hasport cable conversion kit M96-CC. Japanese or American market Y21, Y80, S80, S4C hydraulic clutch transmissions.

Mounts: A combination of stock Civic and Integra mounts can be used if a hydraulic B-series tranny is used. You will need to change the bracket under the timing belt and on the rear of the engine for ones of the appropriate year or Hasport mount kit with cable conversion included M92-B16-10

ECU: Using non-OBD ECU is not a good idea. A PR3 from 92-93 Integra RSi or XSi, a P30 or modified P28 for B16A a P61 or modified P28 for B17A or a 94-95 PR4 for B18A are acceptable.

Electrical differences: CX and VX need to convert to 4 wire O2. Non VTEC models need VTEC solenoid and VTEC pressure. All models modify stock harness to fit, add knock sensor except when using chipped P28.

Axles: 90-93 Integra axles, 94 up Integra axles, 94-97 Del Sol VTEC axles or 99-00 Civic Si axles. Note: The intermediate shaft seal in the left side of the transmission is different for 90-93 Integra axles. Depending on axle transmission combo, it may have to be changed.

Shift linkage: 94 up Integra linkage for Civics. 94-97 Del Sol VTEC linkage for Del Sol

Upper radiator hose: For B16A and B17A, use 94 Del Sol VTEC. For B18A use 94 Integra LS

Lower radiator hose: 94 Del Sol VTEC

Throttle cable: B16A and B17A use 92 GSR; B18A use 90 Integra LS

Clutch hydraulics: Stock

Air conditioning: 94-95 Del Sol VTEC AC bracket.

Special instructions: For 1st gen B16As, replace distributor with one from 92-93 GSR, 94-95 GSR or 94-95 Del Sol VTEC. For 90-91 B18As replace distributor with one from 92-95 Integra B18A or B18B. Injectors need to be replaced in both engines with ones from above models or from 92-95 Civic EX, Si, DX or LX.

Special considerations: The first gen B16A engines are cheap, best bang for the buck. 92-93 B18As and GSRs have compatible electronics and make for an easy swap.

Swap Info 5th Generation1992-1995 Civic and Civic Del Sol with 1994-2000 Integra LS B18B Engine

Transmission: These engines come with hydraulic clutch transmissions including Japanese or American market Y21, Y80, S80, and S4C. Japanese or American market S1, J1, A1, Y1 or YS1 cable clutch transmissions w/ Hasport cable conversion kit M92-CC can be used.

Mounts: A combination of stock Civic and Integra mounts can be used if a hydraulic B-series tranny is used. On some 96 and up engines you may need to change the bracket under the timing belt and on the rear of the engine for ones of the appropriate year.

For cable clutch B-series transmissions, used Hasport cable hydraulic conversion kit M92-CC

ECU: PR4 from 94-95 Integra

Electrical differences: CX and VX need to convert to 4 wire O2. All models modify stock harness to fit.

Axles: 90-93 Integra axles, 94 up Integra axles, 94-97 Del Sol VTEC axles or 99-00 Civic Si axles. Note: The intermediate shaft seal in the left side of the transmission is different for 90-93 Integra axles. Depending on axle transmission combo, it may have to be changed.

Shift linkage: 94 up Integra linkage for Civics. 94-97 Del Sol VTEC linkage for Del Sol

Upper radiator hose: 94 Integra LS

Lower radiator hose: 94 Del Sol VTEC

Throttle cable: 94 Integra LS

Clutch hydraulics: Stock

Air conditioning: 94-95 Del Sol VTEC AC bracket.

Special instructions: Using OBD II ECU requires involved wiring modifications. Generally, if the ECU you choose does not use the same plugs as the stock ECU, an adapter harness will cut down on a lot of potential problems.

Special considerations: Nice motor, smog legal, and excellent base for turbo car. Smog legal in CA.

Swap Info 5th Generation 1992-1995 Civic and Civic Del Sol with 1994-2001 Integra GS-R B18C (including Integra Type R) & 1993-1997 Del sol VTEC and Civic Si B16A Engine

Transmission: These engines come with hydraulic clutch transmissions including Japanese or American market Y21, Y80, S80, and S4C. Japanese or American market S1, J1, A1, Y1 or YS1 cable clutch transmissions w/ Hasport cable conversion kit M92-CC can be used.

Mounts: A combination of stock Civic and Integra mounts can be used if a hydraulic B-series tranny is used. On some 96 and up engines you may need to change the bracket under the timing belt and on the rear of the engine for ones of the appropriate year.

For cable clutch B-series transmissions, use Hasport cable hydraulic conversion kit M92-CC

ECU: Use ECU that came with the engine with proper modification to wiring. LS-VTEC and CR-VTEC should use modified VTEC ECUs. For 96 up engines either back date electronics and use appropriate OBD I ECU or use adapter harness and make proper wiring modifications.

Electrical differences: For B18C add air intake bypass wiring. CX and VX need to convert to 4 wire O2. Non-VTEC models need VTEC solenoid and VTEC pressure. All models modify stock harness to fit, add knock sensor except when using chipped P28.

Axles: 90-93 Integra axles, 94 up Integra axles, 94-97 Del Sol VTEC axles or 99-00 Civic Si axles. Note: The intermediate shaft seal in the left side of the transmission is different for 90-93 Integra axles. Depending on axle transmission combo, it may have to be changed.

Shift linkage: 94 up Integra linkage for Civics. 94-97 Del Sol VTEC linkage for Del Sol

Upper radiator hose: For engines with B16A head or Type R, use 94 Del Sol VTEC. For engines with B18C head, use 94 GSR

Lower radiator hose: 94 Del Sol VTEC

Throttle cable: For engines with B16A head, use 94 Del Sol VTEC. For engines with B18C head, use 94 GSR. For engines with Type R head, use 98 Type R.

Clutch hydraulics: Stock

Air conditioning: 94-95 Del Sol VTEC AC bracket.

Special instructions: Using OBD II ECU requires involved wiring modifications. Generally, if the ECU you choose does not use the same plugs as the stock ECU, an adapter harness will cut down on a lot of potential problems.

Special considerations: US motors are smog legal as long as the motor is from the same or later year vehicle. Japanese type R motors are very potent and wiring is relatively easy with ECU adapter harness. The B series engine swaps into these year vehicles are by far the easiest. There are Honda mounts available to bolt those engines directly in and the wiring is by far the simplest. The majority of people opt to put in a B18C, B18B or 94-95 Del Sol B16A and those bolt in directly using Del Sol VTEC mounts. The Del Sol mounts are the same as the 94 up Integra mounts with the exception of the AC bracket.

Swap Info 5th Generation 1992-1995 Civic and Civic Del Sol with 1994-2000 H22A, H23A Prelude and 1996-2000 F22 Accord Engine

Transmission: All 92-up Prelude 5-speed transmissions and 92-up Accord 5-speed transmissions. Transmissions with ATTS will require major wiring modifications.

Mounts: Hasport mount kit M92-H22-10

ECU: Use ECU that came with the engine with proper modification to wiring. For 96 up engines either back date electronics and use appropriate OBD I ECU or use adapter harness and make proper wiring modifications.

Electrical differences: CX and VX need to convert to 4 wire O_2. Non-VTEC models need VTEC solenoid and VTEC pressure. Add EGR solenoid and EGR valve except for VX. All models modify stock harness to fit, add air intake bypass wiring, and add knock sensor.

Axles: Use 90 up Accord Intermediate shaft with 90-93 Integra axles. To use the Prelude intermediate shaft a shorter left-hand axle with a female inner joint must be used. To construct the proper left-hand axle use 94 GSR left-hand axle inner and outer joints on an 86-89 Integra axle shaft.

Shift linkage: These transmissions use cables instead of linkage. The shifter mechanism and cables from these vehicles may be used, 90 up Accord 5spd or 92-up Prelude 5spd cables and linkage. This will require some cutting and drilling.

Upper radiator hose: 90 Integra LS, trim to fit.

Lower radiator hose: 94 Del Sol VTEC

Throttle cable: 98 Prelude

Clutch hydraulics: Stock

Air conditioning: Not without some custom fabrication.

Special instructions: The mounting bracket on the right-hand frame rail will need to be removed or it will interfere with the shifter assembly. A thin fan or pusher fan will have to be used with the radiator. Radiator should be upgraded to a Del Sol VTEC radiator for increased cooling.

Special considerations: Suspension and brake upgrades a good idea. The H22A is 175 to 185 lbs heavier than stock D-series drive trains and 85 lbs heavier than a B16A drivetrain

THE SIXTH GENERATION, EK HONDA CIVIC 1996-2000, CX/DX/HX/EX, SiR, Si (Canada)

The EK Civic is now also gaining in popularity from the modification crowd. Although its chunky looks are not as sleek as the EG, it still has large, nice-looking projector headlights and sleek lines.

Due to stricter collision and side impact standards that came into effect in the US during the car's tenure, the EK is more heavily reinforced and thus not quite as light as the EG. Although this weight makes the car a bit slower, it is stronger in case of an accident.

The EK still has Honda's superior multilink suspension and can be modified into a G machine. With a good motor and the right suspension brake and tire mods, this combo is still right in there with some of the world's best handling cars.

Like the EG, the EK can receive a workhorse B series motor as a straight bolt in, with the exception of changing some factory mounts. Because the EG falls into the years of production when OBDII came into effect, there are a few minor complications as far as wiring issues go. The 1997–2000 Civic Si came with a B16A motor as standard equipment, which makes this swap even more straightforward. When building an EK Hybrid, it is a good idea to swap to an OBDII version of the B series motor that you are dropping in. It is also a good idea to install the complete harness and ECU from the same year and type as the motor to avoid OBDII problems like poor running and chronic check engine lights.

With work and a Hasport swap kit the EK can also get motivation from an H22 or H23 Prelude engine. Even the F22 Accord engine can be made to fit with a proper kit.

Swap Info 6th Generation 1996-2000 Civic and 1996-1997 Del Sol with 1st generation 1989-1991 JDM Civic B16A, 1992-1993 Integra GSR B17A, and 1990-1993 Integra B18A Engine

Transmission: These engines come with Japanese or American market cable clutch transmissions including S1, J1, A1, Y1 or YS1. To use them, you will need the Hasport cable conversion kit M96-CC. Japanese or American market Y21, Y80, S80, S4C hydraulic clutch transmissions.

Mounts: For Civic, a combination of stock 99 Civic Si and Integra or Del Sol VTEC mounts and brackets can be used if a hydraulic B-series tranny is used. For Del Sol the Del Sol VTEC mounts and brackets can be used and you will need to change the bracket under the timing belt and on the rear of the engine for ones of the appropriate year. When using B-series engines with cable clutch transmissions, you will need to change the bracket under the timing belt and on the rear of the engine for ones of the appropriate year and use Hasport cable conversion included M96-CC.

ECU: By far the easiest ECU to use in most situations is the P30 from the 96-97 Del Sol VTEC. Only a few models of 99-00 Civics need the 99-00 Si ECU. For more info see Electrical Differences below.

Electrical differences: Wiring differences can be significant; here are a couple of guidelines. The electrical modifications depend entirely on the ECU used and the year and model of the vehicle. The engine electronics need to be updated to match the year of the ECU used. For the engine wiring harness, it is generally best to use the stock harness and modify it to fit. If the ECU you choose does not use the same plugs as the stock ECU, an adapter harness will cut down on a lot of potential problems.

Axles: 90-93 Integra axles, 94 up Integra axles, 94-97 Del Sol VTEC axles or 99-00 Civic Si axles. Note: The intermediate shaft seal in the left side of the transmission is different for 90-93 Integra axles. Depending on axle transmission combo, it may have to be changed.

Shift linkage: 99-00 Civic Si linkage for Civics. 94-97 Del Sol VTEC linkage for Del Sols

Upper radiator hose: For B16A and B17A, use 94 Del Sol VTEC. For B18A use 94 Integra LS

Lower radiator hose: 94 Del Sol VTEC

Throttle cable: B16A and B17A use 92 GSR. B18A use 90 Integra LS

Clutch hydraulics: Stock

Air conditioning: 96-97 Del Sol VTEC AC bracket.

Special instructions: Using non-OBD ECU is not a good idea. Using OBD I ECU requires the use of either ECU adapter or involved modifications to the harness.

Special considerations: Best bang for the buck, but not going to be smog legal in CA.

Swap Info 6th Generation 1996-2000 Civic and 1996-1997 Del Sol with 1994-1995 Integra LS B18B 1996-2001 Integra LS B18B (also LS-VTEC CRV-VTEC) Engine

Transmission: These engines come with hydraulic clutch transmissions including Japanese or American market Y21, Y80, S80, and S4C. Japanese or American market S1, J1, A1, Y1 or YS1 cable clutch transmissions w/ Hasport cable conversion kit M96-CC can be used.
Mounts: A combination of stock Civic and Integra mounts can be used if a hydraulic B-series tranny is used. On some 96 and up engines you may need to change the bracket under the timing belt and on the rear of the engine for ones of the appropriate year.
For cable clutch B-series transmissions, use Hasport cable hydraulic conversion kit M96-CC
ECU: 96-up PR4 from 96-00 Integra or 94-95 PR4 from 94-95 Integra with ECU adapter plugs.
Electrical differences: Wiring differences can be significant; here are a couple of guidelines. The electrical modifications depend entirely on the ECU used and the year and model of the vehicle. The engine electronics need to be updated to match the year of the ECU used. For the engine wiring harness, it is generally best to use the stock harness and modify it to fit. If the ECU you choose does not use the same plugs as the stock ECU, an adapter harness will cut down on a lot of potential problems.
Axles: 90-93 Integra axles, 94-up Integra axles, 94-97 Del Sol VTEC axles or 99-00 Civic Si axles. Note: The intermediate shaft seal in the left side of the transmission is different for 90-93 Integra axles. Depending on axle transmission combo, it may have to be changed.
Shift linkage: 99-00 Civic Si linkage for Civics 94-97 Del Sol VTEC linkage for Del Sols
Upper radiator hose: 94 Integra LS
Lower radiator hose: 94 Del Sol VTEC
Throttle cable: 94 Integra LS
Clutch hydraulics: Stock
Air conditioning: 96-97 Del Sol VTEC AC bracket.
Special instructions: Using OBD I ECU requires the use of either ECU adapter or involved modifications to the harness.
Special considerations: Nice motor, and excellent base for turbo car. Smog legal in CA.

Swap Info 6th Generation 1996-2000 Civic and 1996-1997 Del Sol with 1996-1997 Del Sol Vtec or 1996-2000 Civic Si B16A and 1994 to 2001 Integra GS-R B18C and B18C5 incl.Type R (also LS-VTEC CRV-VTEC) Engine

Transmission: These engines come with hydraulic clutch transmissions including Japanese or American market Y21, Y80, S80, and S4C. Japanese or American market S1, J1, A1, Y1 or YS1 cable clutch transmissions w/ Hasport cable conversion kit M96-CC can be used.
Mounts: A combination of stock Civic and Integra mounts can be used if a hydraulic B-series tranny is used. On some 96 and up engines you may need to change the bracket under the timing belt and on the rear of the engine for ones of the appropriate year. For cable clutch B-series transmissions, used Hasport cable hydraulic conversion kit M96-CC
ECU: Use ECU that came with the engine with proper modification to wiring. LS-VTEC and CR-VTEC should use modified VTEC ECUs. For 92-95 engines either update electronics and use appropriate OBD II ECU or use adapter harness and make proper wiring modifications.
Electrical differences: Wiring differences can be significant; here are a couple of guidelines. The electrical modifications depend entirely on the ECU used and the year and model of the vehicle. The engine electronics need to be updated to match the year of the ECU used. For the engine wiring harness, it is generally best to use the stock harness and modify it to fit. If the ECU you choose does not use the same plugs as the stock ECU, an adapter harness will cut down on a lot of potential problems.
Axles: 90-93 Integra axles, 94 up Integra axles, 94-97 Del Sol VTEC axles or 99-00 Civic Si axles. Note: The intermediate shaft seal in the left side of the transmission is different for 90-93 Integra axles. Depending on axle transmission combo, it may have to be changed.
Shift linkage: 99-00 Civic Si linkage for Civics. 94-97 Del Sol VTEC linkage for Del Sols
Upper radiator hose: For engines with B16A head or Type R, use 94 Del Sol VTEC. For engines with B18C head, use 94 GSR
Lower radiator hose: 94 Del Sol VTEC
Throttle cable: For engines with B16A head, use 94 Del Sol VTEC or 99-00 Civic Si. For engines with B18C head, use 94 GSR. For engines with Type R head, use 98 Type R.
Clutch hydraulics: Stock
Air conditioning: 96-97 Del Sol VTEC AC bracket.
Special instructions: Using OBD I ECU requires the use of either ECU adapter or involved modifications to the harness.
Special considerations: US motors are smog legal as long as the motor is from the same or later year vehicle. Japanese-type R motors are very potent and wiring is relatively easy with ECU adapter harness. The B-series engine swaps into these year vehicles are by far the easiest. There are Honda mounts available to bolt those engines directly in and the wiring is by far the simplest. The majority of people opt to put in a B18C, B18B or 94-95 Del Sol B16A and those bolt in directly using Del Sol VTEC mounts. The Del Sol mounts are the same as the 94-up Integra mounts with the exception of the AC bracket.

Swap Info 6th Generation 1996-2000 Civic and 1996-1997 Del Sol with 1996-2000 Prelude H22A1or H23A1 and 1996-2000 F22 Accord Engine

Transmission: All 92 up Prelude 5spd transmissions and 92 up Accord 5spd transmissions. Transmissions with ATTS will require major wiring modifications.
Mounts: Hasport mount kit M96-H22-10
ECU: Use ECU that came with the engine with proper modification to wiring. For 92-95 engines either up date electronics and use appropriate OBD II ECU or use adapter harness and make proper wiring modifications.
Electrical differences: Wiring differences can be significant; here are a couple of guidelines. The electrical modifications depend entirely on the ECU used and the year and model of the vehicle. The engine electronics need to be updated to match the year of the ECU used. For the engine wiring harness, it is generally best to use the stock harness and modify it to fit. If the ECU you choose does not use the same plugs as the stock ECU, an adapter harness will cut down on a lot of potential problems.
Axles: Use 90-up Accord Intermediate shaft with 90-93 Integra axles. To use the Prelude intermediate shaft a shorter left-hand axle with a female inner joint must be used. To construct the proper left-hand axle use 94 GSR left-hand axle inner and outer joints on an 86-89 Integra axle shaft.
Shift linkage: These transmissions use cables instead of linkage. The shifter mechanism and cables from these vehicles may be used, 90-up Accord 5spd or 92 up Prelude 5spd cables and linkage. This will require some cutting and drilling.
Upper radiator hose: 90 Integra LS, trim to fit.
Lower radiator hose: 94 Del Sol VTEC
Throttle cable: 98 Prelude
Clutch hydraulics: Stock
Air conditioning: Not without some custom fabrication.
Special instructions: A thin fan or pusher fan will have to be used with the radiator. Radiator should be upgraded to a Del Sol VTEC or 99-00 Civic Si radiators for increased cooling.
Special considerations: Suspension and brake upgrades a good idea. The H22A is 175 to 185 lbs heavier than stock D-series drivetrains and 85 lbs heavier than a B16A drivetrain.

THE FUTURE

Although it is too soon to feature the details of such swaps, you can look forward to some new hybrid combinations in the near future. Swapping the new 200 hp K20A2 RSX Type S motor for the 160 hp K20A3 found in the base RSX and Civic Si is a no-brainer.

Getting rid of the anemic D17A motors found in the Civic coupe and swapping in a K20A2 or K20A3 motor with a 6-speed tranny will probably come next. The ingenuity of the Honda crowd will no doubt continue into the next millennium.

6 HOT STREET COMBOS

One of the most appealing things about Honda motors is the plethora of hot parts that can be found right from the pages of the factory parts catalog. The other appealing thing is that you can build a really hot naturally aspirated street motor with some of these relatively inexpensive factory parts. In this chapter we will focus primarily on the hot combinations of JDM (Japanese Domestic Market) and other stock Honda parts that can be used to build a potent street machine. Even though they are stock parts, stock factory parts have a lot of extensive engineering and testing behind them. This makes for superior function and reliability. When used within their design parameters factory stock parts will provide a long and reliable life. With Hondas there is quite a bit of interchange that can be done between years and even models of different motors to come up with some potent off-the-shelf combinations.

It is possible to come up with streetable combinations that belt out over 220 crank and 190 wheel hp with over 150 lbs-ft. of wheel torque. When we list power figures here, it is wheel power confirmed on the Dynojet 248C chassis dynamometer. Usually the crank hp is about 15% greater than these wheel hp figures due to frictional losses in the drivetrain. When mixing and matching the correct factory parts with headwork, headers, intake systems and the usual bolt-ons, one can obtain these power levels fairly inexpensively. These are impressive power numbers and are done on pump gas with clean emissions; good low speed driveabilty, and a silky smooth idle thanks to Honda's amazing VTEC variable cam system.

The number of possible combinations that can be done is mind boggling and some are more effective than others. We will cover some popular swaps and parts that will get some muscle out of your motor. We have not personally tried all of these, as it is impossible to have done all of the possible combinations. Most of the suggestions listed here have been fairly well-documented.

Of course, when assembling new motor combinations, you must "cc" the combustion chamber and piston domes, calculate compression ratios and check piston-to-valve clearances and piston-to-head clearance. These are good practices when assembling any motor but it is even more critical when putting together combinations that were not designed to be run with each other. Also, remember that headwork, combustion chamber cloverleafing, head milling, adjustable timing gears, changing cams and pistons can all affect compression ratio. Too high of a compression ratio can result in deadly detonation. So be careful, measure everything, and have fun.

You might want to review our prior sections on the cylinder head, camshafts and the bottom end to understand some of the theory behind some of these parts. That information can make your choices below easier to understand as we will only be getting into general theory here.

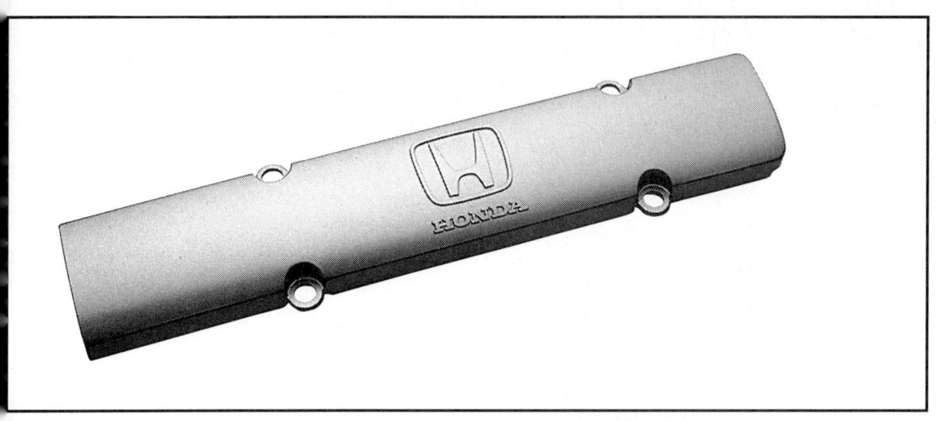

Not all JDM parts are strictly for speed. SPW sells these trick-looking JDM valve covers and plug wire covers for dress up purposes.

FRANKENSTEIN MOTORS

Perhaps the biggest bang for the buck is a "Frankenstein Motor." A Frank is a bunch of swaps between different blocks and heads within the same engine family to make some non-factory combinations that work well.

The most popular Frank swap is adding a VTEC B16A or B18C head from a Del Sol, a Civic Si or an Integra GSR to an LS, non-VTEC Integra B18B or a CRV B20B or B20Z bottom end. The common LS Integra B18A and B18B engines have a bore and stroke of 81x89mm, which gives you 1834cc. The powerful but expensive VTEC B18C's have a 81x87.2 bore and stroke which gives you 1797cc of displacement. The smaller but powerful B16A has a bore and stroke of 81x77.4mm for a displacement of 1595cc. By putting the VTEC head on the LS bottom end you pick up 35ccs over the B18C and a whopping 239cc more torque-producing displacement over the B16A. The added displacement and stroke give the LS Frank motor a nice torque advantage.

VTEC Head to LS Bottom

Due to Honda's excellent interchangeability, you can put a B16A head, a B18C head or even the expensive and rare B18C5 head from an Integra Type R on the LS bottom end to make a larger displacement, longer stroke, potent torque monster B18. It has the best of both worlds, the torque of the B18B or B18A with the screaming VTEC power of the B18C. This head swap adds about 40 hp to your typical LS motor with naturally aspirated, horsepower figures in the 170–190 hp range and torque in the 120–145 lbs-ft. zone. The engine will retain a compression ratio of approximately 10:1 with this combination of heads.

To add a VTEC cylinder head to an LS bottom end, you must tap and plug the VTEC oil supply hole found on the bottom left side of the head. A 1/8" pipe plug will work well for this. Next you must run an oil supply line from a T-fitting placed on the oil pressure sending unit boss on the block to the VTEC oil galley plug found on the intake side of the head near the distributor. To make the oil line, you need a piece of –4 braided steel line about 20" long, with two female A/N fittings attached (an industrial hydraulics shop can make this for you), a 3/8" NPT to –4A/N adaptor for the head, a 1/8" NPT to –4

A/N adaptor and a 1/8" NPT T-fitting with two female sides and one male side.

You must also open up the dowel pin holes on the VTEC cylinder head to 9/16 of an inch as the LS block has larger dowel pins to locate the head on the block. This is a pretty simple operation that can be preformed on a drill press with care. If you don't feel like doing this, the operation can be easily handled by any competent machine shop.

An LS head gasket is used with the appropriate VTEC ECU for your year and chassis of vehicle.

Wiring—A jumper wire running from the appropriate pin in the ECU to the VTEC control solenoid completes the VTEC activation. The proper VTEC ECU is the ECU for the VTEC model of your vehicle. See chart above for wiring help.

It is also critical for you to hook up the knock sensor or the VTEC function will not work. The engine's ECU looks for the knock sensor signal in order to activate VTEC. The B18A and B18B block has no provision for a knock sensor. In this case or if you don't want to run a knock sensor, you can do several things. You can have a knock sensor not attached but grounded to the chassis hooked up to your ECU or Hasport has ECUs that are reprogrammed to allow VTEC operation with no knock sensor input.

Year	VTEC Solenoid	VTEC Pressure Pressure	Knock Sensor
1988-91	A-8	B-5	B-19
1992-95	A-4	D-6	D-3
1996-00	A-8	C-15	D-6, 99-00 Si C-3

Here is a chart with the pin locations where the wires to control the VTEC solenoid connect.

B20 Bottom and VTEC Head

Another very potent Frankenstein combination is using the B20 bottom end from the CRV mini sport utility, with a VTEC cylinder head. The B20 bottom ends that are desirable are the B20B (97-98 CRV) and the B20Z (99-01 CRV). With a big bore of 84mm and a stroke of 89mm, with a whooping displacement of 1973cc, a B20 Frank has the potential to be the meanest of all the Frank motors.

Of the two B20 motors, the B20Z has more compression at 10.2:1 over the B20B's lower 8.8:1. This is because the B20Z has a flatter top piston when compared to the B20B's deeper dish. The lower compression B20B is more desirable for a bolt-on supercharger or turbocharger kit because of this. If you wanted to run high compression with a B20B/Z, there are no factory pistons available, but any number of quality custom forged piston makers, such as JE, Aires or Wisco, can make a piston for the B20B/Z. This piston should be made with a dome volume to give compression ratios from 11:1 for the street to 13:1 for race gas. With 11:1 pistons and a VTEC head mildly worked over, it can be possible to have a very impressive motor for very little money.

The procedure to swap a VTEC head on the B20B is the same as the LS swap with the exception of using the B20 head gasket. A naturally aspirated, mild B20B can easily get over 200 hp on pump gas with over 150 lbs-ft. of stump-pulling torque.

Check Clearances—As a warning, when the VTEC head is installed, there is not a whole lot of piston-to-valve clearance as VTEC heads have larger 33mm intake valves vs. the B20 motors 31mm valves. If you are dyno tuning, you should avoid advancing the intake cam much as contact will occur between the piston and valves if the cam is advanced much more than 4 degrees. In fact this is so close, it is not really advisable to advance the intake cam at all.

If running larger lift and duration aftermarket cams in your B20/VTEC Frank, or if the head has been modified for higher compression by milling, the piston's valve pockets should be modified and the valve-to-piston clearance confirmed with clay before the head is final assembled to the block. Additional clearance can be obtained by careful grinding the pistons valve reliefs with a die grinder with carbide burrs and polishing the result with cartridge rolls. A minimum clearance of 0.045" on the intake valves and 0.055 on the exhaust valves is advisable.

H & D Motors

For H motors, running the H22 VTEC cylinder head on the Non-VTEC H23 bottom end is an easy way to gain 100cc of power adding displacement. The procedure to do the switch is very similar to the LS VTEC or B20 VTEC swaps. Another way is to add the H23 crank, rods and pistons to the H22 block. The Accord F22 crank, rods and pistons in the H22

JDM Civic Type R camshafts from a JDM B16AS are pretty potent and are a good upgrade.

block can also be used to make a 2300cc motor.

There are even some swaps that you can do to get more out of your D16. The best head to use on the D16 is the 1996-and-up single cam VTEC head D16Y8. This head had pentroof-style combustion chambers with generous quench zones. The quench areas help improve combustion, reducing the need for as much advanced timing and reduces the propensity to detonate. The 1992–1995 SOHC VTEC head had a circular combustion chamber that does not work as well. The newer head also raises the compression ratio by a half point.

CAMSHAFTS AND VALVETRAIN

With the B-series engine, you are in luck. Between the VTEC B-series motors, you can swap cams back and forth without any problems. The hot cams to get in this case are the USA market Type R Integra cams or the JDM (Japanese Domestic Market) Civic Type R cams.

These factory cams are quite respectable, with near racecar lift, duration and overlap on their high rpm lobes. Being VTEC, they still purr like a stock motor at idle and low speeds and will still be able to pass your local smog test.

With the exception of the Type R cams, all of the standard B-series motors have similar high rpm VTEC lobe specs. They all have 230 degrees of intake duration (measured at 1mm of checking clearance) with 10.6–10.7mm of lift. On the exhaust side, they have 227 degrees of duration and 9.4 mm of lift. The cams all have about 17 degrees of overlap.

Type R Cams

What works really well on all of these motors is to swap the cams for the Type R Integra cams or the JDM 1998 Civic Type R cams. The Integra Type R cams have 240 degrees of intake duration with 11.5 mm of lift while the exhaust sports 235 degrees of duration and 10.5 mm of lift. The Integra Type R cam has 25 degrees of overlap. The JDM 1998 Civic Type R cams are slightly bigger, with 243 degrees of intake duration and 28 degrees of overlap.

When installing these cams on a non-Type R, VTEC B-series motor, you should put the stock B-series valve springs on the exhaust side of the head and get the appropriate Type R intake valve springs. US-market B motors, with the exception of the

TYPE R CAMS & SPRINGS

Type-R Part	US B18C5 Type R	JDM B16B 1998 Civic Type R
Intake Cam	14111-P73-J00	14111-PCT-000
Exhaust Cam	14121-P73-J00	Same
Inner Valve Spring	14751-P73-J01	Same
Outer Valve Spring	14761-P73-J01	Same

Here is a chart with the part numbers of the hot Type-R cams and the valve springs needed to run them, at least on the exhaust side.

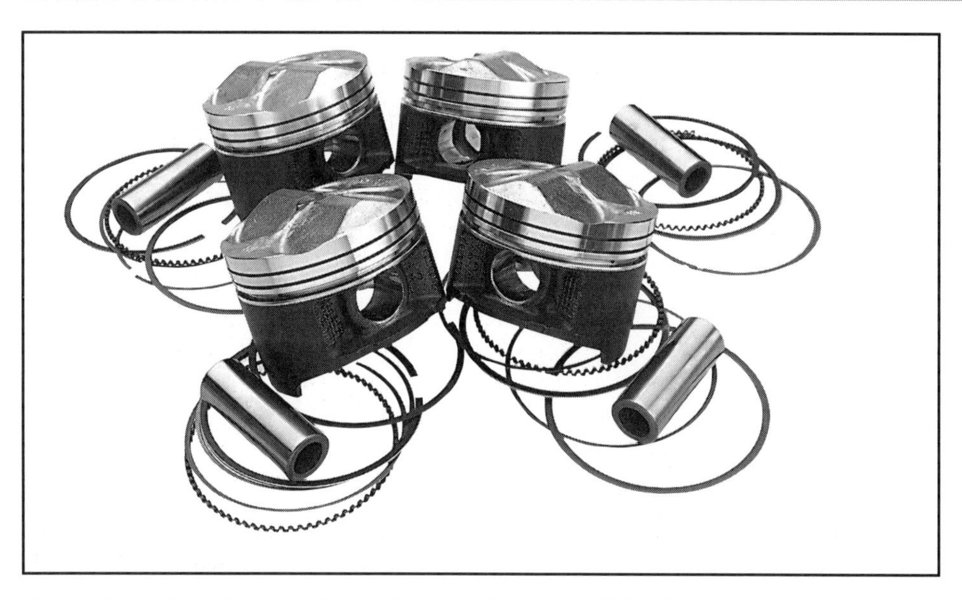

The black dry film lubricant skirt coating can identify the JDM ITR pistons. This helps reduce friction and wear. The coating is very similar to the expensive aftermarket coatings that many racers have applied to their pistons.

Type R, only have dual springs on the intake valve. The dual exhaust valve spring is important to help handle the extra valvetrain velocity that the much higher lift Type R exhaust cam has. The regular B series dual intake valve spring setup can be run with the Type R cams because they are actually stiffer than the Type R intakes. This is because the Type R valves are lighter than the regular B-series valves.

Type R cams can gain from 8–9 more top-end horsepower, sometimes even more when the cam timing is optimized, over the stock B series cams and are an excellent value in streetable performance cams.

As a cost-saving measure, to avoid having to buy exhaust valve springs, some people are having success installing only the Type-R intake cam. Just the intake cam alone can result in gains from 6-7 hp.

Type R Valves

You may also want to run the Type-R valves. Although the Type R valves are the same diameter as the regular B series valves, they have an improved contour for better flow and they are lighter. A good head porter can contour the stock B series valves to match the Type-R valves, but if you need new valves, the Type-R valves are preferable.

Type S Cam For H Series

For VTEC H-series engine lovers, there is the JDM Prelude Type S cam. The Type S cam has more duration and lift than the standard US market VTEC cam. A H22 put together with a few JDM Type S parts can put out 185–195 wheel hp with wheel torque readings in the 150–155 lbs-ft. range.

Aftermarket Cams

If you don't want to run the Type R cams, desiring even more top end power, excellent streetable cams are made by Toda, Spoon Sports, Skunk 2 and JUN. The big American companies like Crane and Crower are also getting into the game with their own billet offerings. When running other cams, please note that on finger follower motors like Honda's, it is critical to run near-stock base circle billet cams for long life and proper true-to-spec valvetrain geometry. Regrind cams just don't work well with these motors. Poor peaky powerbands and rapid wear can result from regrinds. There are not too many billet offerings for the D series motors currently. Comp Cams and Crane Cams are about the only current offering and JUN is planning billet offerings in the near future.

With aftermarket cams, it is also important to run the manufacturer's recommended, compatible valvetrain parts, most importantly the springs and retainers. See Chapter 2, pages 33-43 for more technical details on cams.

PISTONS

For an all-motor, street-driven car, the factory Type R pistons are the best bet. This is the US or JDM Integra Type R piston for the B18C and the LS VTEC Frank motor with the JDM Civic Type R piston for the B16A.

Type R pistons are a high-pressure, die cast construction. This is the best way to make a cast piston. Since they are cast, they are not the best choice for a nitrous-burning motor but they are excellent in all motor applications. An advantage for cast pistons in a daily driver sort of car is that they can use a much tighter piston-to-wall clearance. This makes for an engine that is quieter and burns less oil. Cast pistons are also easier on the cylinder walls.

Type R pistons have a black, dry, film lubricant coating on the skirts for longer wear and less friction as well as additional oil holes in the ring grooves for better high rpm oil control. Type-R pistons also have holes in the wrist pin bosses to improve wrist pin oiling. As an interesting and useful fact, all of the B-series motors have the same compression height, so the pistons interchange.

Compression Ratios

For a Honda street motor that will run on pump gas, the maximum compression that should be run is in the low 11:1 range. The reason why Hondas can get away with this high compression ratio with modern unleaded fuel is mostly because of their superior combustion chamber shape and small bore diameters. Higher compression than this requires racing gas to avoid detonation. With cast pistons, detonation should be avoided at all costs as cast pistons are more brittle than forged pistons.

A set of H22 Type S pistons has a higher dome and greater compression than the stock pistons. The dome is good for over a point of compression.

Type R Pistons

The Integra Type-R piston is available in two versions: the P73-00 JDM version and the P73-A0 US market version. The difference between the two is that the JDM piston has a slightly taller dome, which gives about 0.2 of a point higher compression. The US Type-R A0 piston is the piston to use in the larger displacement LS Frank motor, as it will yield approximately 11:1 compression with this combo due to its lower dome. The JDM 00 piston is the piston of choice for a B18C as it will yield about 11:1 in this combo.

The JDM Civic Type R or PCT piston has the highest dome of all the Type-R pistons to get 11:1 compression out of the smaller B16 motor. It is not recommended that this piston be installed in a B18C or a B18B or a Frank motor as the compression ratio will end up being unstreetable, close to 12:1. This compression is good for mild race application with racing fuel and is in fact a popular combo for Hybrid Civics raced in NASA's PTTC road racing class.

Prep—When running Type-R pistons on a B16A or a B17A motor, you must remove 1mm of material on each side of the small end of the connecting rod in order to have sufficient clearance where the rod meets the piston. Any automotive machine shop can easily do this operation.

As a warning, since 11:1 is close to the limit of streetable pump gas compression and the fact that these pistons have relatively high domes, it is important to verify your engine's compression ratio, valve-to-piston and piston-to-head clearance if your head has been modified by milling or quench welding before final assembly. The head and piston domes should be cc'd and the clearances checked by the clay method to ensure that no interference will occur.

Type S Pistons

For the H22, JDM Prelude Type S piston will raise the engine's compression into the high 10:1 range and are a direct replacement. The Type S pistons do not have the anti-friction coating of the Type-R pistons but have a higher volume dome. For the SOHC D16 motor, installing the pistons from a 1986–1989 Integra motor bumps the compression about 3/4ths of a point. If you have a VTEC D motor and swap to a 96-and-later head, this, combined with the Integra pistons, will get you a respectable compression ratio in the high 10:1 range.

CONNECTING RODS

The JDM Civic Type R with the B16A motor has improved connecting rods featuring steel with higher chromium content. Adding chromium to the steel greatly increases its strength; much like the chromium and molybdenum in chrome-moly makes it much stronger than regular steel. This steel is standard in the US B18C and Integra Type-R rods, but not in the US B16A. The B16A has pretty decent rods, however, so it is probably not worth it to go out and buy a set of JDM rods. But if you do have a

The JDM CTR rod features a stronger steel with more chromium in it. This is sort of like Chrome-Molly, the well-known high strength US spec steel. In this picture you can see the stout forged construction, beefy rod bolts and spot faced rod bolt bosses, all features that add strength.

choice, it is good to know that the B16B rods are stronger.

Type S Rods

The H22 Type-S rods are also made of the same stronger steel, but like the B16A, the H22 has pretty decent stock rods, so it is not absolutely necessary to run out and buy a set of new JDM Type-S rods. If you have the choice between the two, grab the Type-S rods and run.

SOHC D16 Rods

The rods on the standard Civic SOHC D16 motor, unlike most Honda rods, are fairly weak parts. The D16 rod suffers from weak bolts and a spindly skinny construction. The stock rods should not be pushed past 200 hp or 7500 rpm, whichever comes first, as they will fail quickly. A quick and dirty junkyard upgrade is to use the rods from a 1986–1987 D16A1 Integra motor. These rods are stronger and can take up to 250 hp. Use the correct Integra bearings with this mod. The rods from the 88–89 version of this motor are just as weak as the D16 ones! The 1985–1987 JDM Civic and CRX motor, the ZC, has a connecting rod that is another bolt in upgrade that is good for about 250 hp. For more strength, there is another junkyard alternative to a racing rod. The B18A, non-VTEC rod can be used with some minor modifications. The small end must be rebushed to take a bushing with a smaller ID as the D16 has a smaller wrist pin, or custom pistons can be made using a B motor wrist pin. The big end of the rod must also be milled thinner to the same thickness as the D-series rod. These operations can be done cheaply by any good automotive machine shop. The modified B18A rod can easily take up to 300 hp and 8500 rpm.

Any stock rod can benefit from polishing the beams and shotpeening. These operations can improve the rod's fatigue strength by over 100%.

INTAKE AND EXHAUST MANIFOLDS

Type R Intake Manifold

A popular swap to gain some additional power is to run an Integra Type-R intake manifold. The Type-R manifold is a short, larger diameter, single-stage manifold with a larger plenum. It is lacking the dual-stage runner system of the B18C motor. The Type-R manifold is optimized for top end power. When doing a manifold swap, you can use either a US Integra Type-R or JDM Civic Type-R manifold. Both work equally well. The only difference is the location of one vacuum fitting on the manifold's plenum, which has absolutely no effect on anything.

Installing On B16A Head—The Type-R manifold is good for about 6 wheel hp at high rpm with perhaps a slight loss in power below 5000 rpm. The manifold is a direct bolt on to the B16A cylinder head and thus works well for any B16A or any Frank motor using this head. It also fits on the rare B17A1 93–99 Integra GSR motor. On the B18C, the flange of the manifold must be extensively modified for it to fit as only the bottom manifold-to-head bolts line up. There is also a problem with the location and alignment of the coolant passage. This must be welded up and ported for the correct shape. The upper bolt holes must be welded up and redrilled in the correct location and the head match-ported to the manifold. After welding, the manifold's head flange must be resurfaced so it will seal properly. After all of this mess, the injectors hand down into the airstream and disrupt flow, making this not the optimal way to do things.

Perhaps a better way is to cut the flange off of both manifolds and weld the B18C flange onto the Type-R manifold. After welding, the manifold's runners can be cleaned up and matched-ported for good flow. Fortunately, if this seems like a great deal of work, Skunk 2 has come up with a bolt-on copy of the Type R manifold that bolts right onto the B18C with no hassles.

To do a clean install of a Type R manifold onto a B16A head, you need these following parts:

- a fuel rail from a 94 Integra
- a 93 B16A Del Sol, a 1999 Civic Si, a 1992-or-later JDM B16A or a B18C motor
- an Idle Air Control Valve and 12mm mounting bolts from a 1992-and-later Integra

The JDM Civic Type R B16A part is a good manifold; it has short runners, good for high rpm power and a larger plenum chamber. It is close to identical to the US Integra Type R intake manifold.

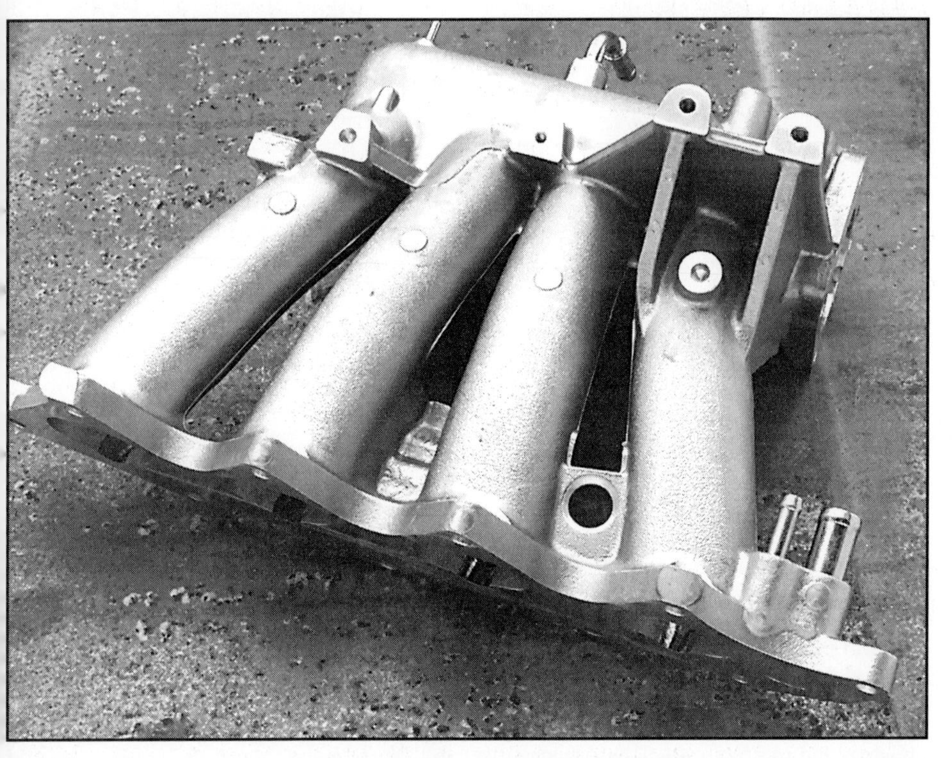

Here is a shot of the bottom of the JDM CTR manifold; you can really see the short, straight runners here.

• a 92-and-later Civic Si or EX or a 97-98 CRV
• a throttle body gasket from an Integra Type-R
• an intake manifold gasket from an Integra Type-R

A 62mm Type-R throttle body, to replace the stock 60mm, is a nice addition at this point also but it may be cheaper to get an aftermarket 64mm throttle body from RC engineering or JG.

JDM Type S Intake Manifold

An H22 motor can benefit from the JDM Type S manifold. This manifold retains the same dual stage design but has bigger diameter and shorter runners, a bypass valve body with bigger ports and a larger plenum chamber. The Type S also has a bigger throttle body. As with the B motors, it might be cheaper to get an aftermarket throttle body rather than shell out the bucks for a brand-new JDM piece.

JDM Integra Type-R Exhaust Manifold

On the exhaust side, a very effective upgrade is to install the JDM Integra Type-R exhaust manifold. This is good for about 6 more wheel hp over the stock manifold. Interestingly enough the JDM manifold is a 4-1, equal length, tubular stainless steel header! The JDM Type-R exhaust manifold fits all B-series motors.

The Type S H22 also has a tubular stainless header but it is of the Tri-Y design, which is good for a broad powerband. The different part of the Type S exhaust manifold is the upper portion of the exhaust manifold. The JDM upper manifold bolts to the stock Prelude downpipe.

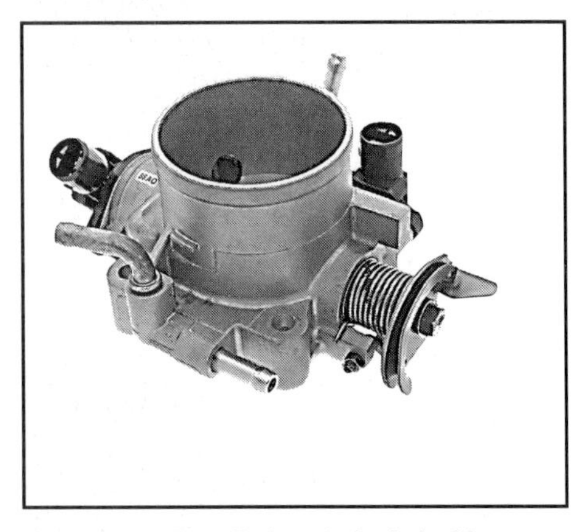

The Integra Type R throttle body is 62 mm up from the stock 60 mm B18C throttle body.

It is impressive that with a well thought-out combination of factory parts, good assembly and some headwork, a Honda motor can obtain very respectable horsepower and torque figures while still keeping factory-like reliability and loan-to-your-grandma driveabilty. There are not too many engine families on earth that can make this claim from any manufacturer. With Honda's cheap and reliable power is just a call to a JDM engine importer or the dealer away.

THE NEAR FUTURE

With Honda's recently released motors, the potential for swapping still exists. Theheads of the RSX Type S 200 hp K20A2 can probably be used to boost the power of the 160 hp K20A3 found in the base RSX and Civic Si. The hotter cams from the S2000 may interchange with the K20 motors. Factory bolt-on performance will continue with the new breed.

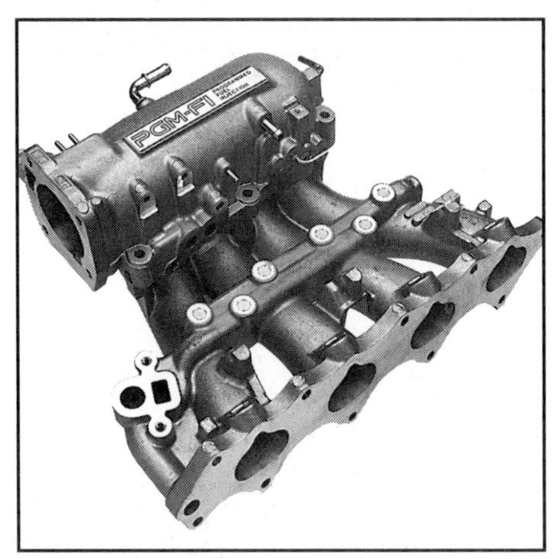

A JDM Prelude Type S manifold is good for 5-6 hp on a USDM H22 motor.

The Type S manifold also has a bigger plenum tank.

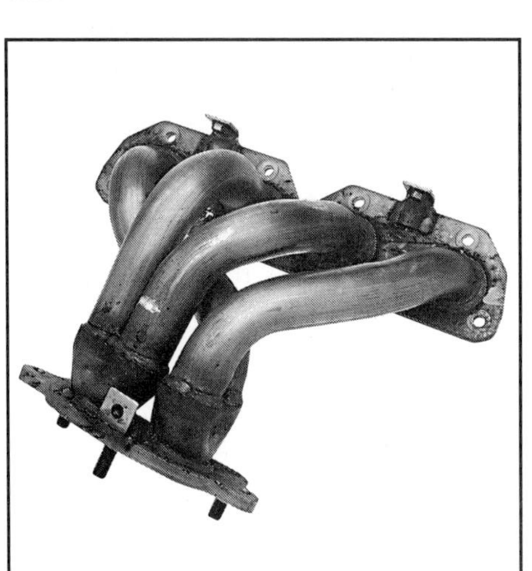

The JDM Prelude Type S manifold is a tubular tri-y header that replaces the stock cast manifold. It bolts to the stock Prelude down pipe. The Type-S manifold is also made of stainless and can retain stock heat shields.

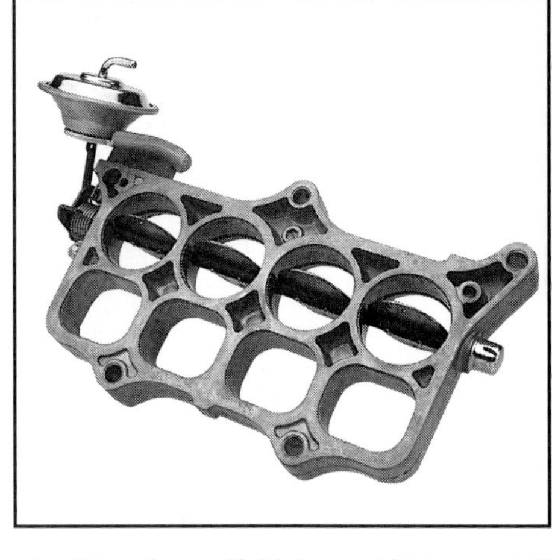

The Type S manifold has a bigger ported bypass valve for more flow.

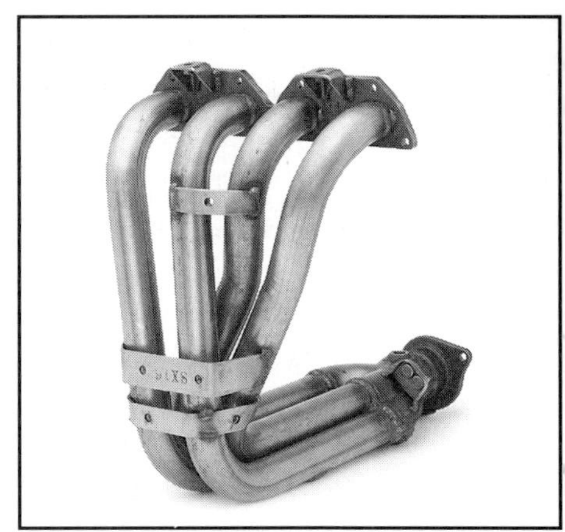

The JDM ITR header works on the B16A and B18C motors, it has a slick 4-1 layout which is good for top end power over the stock tri-y manifolds. The JDM manifold features quality 100% stainless construction. Note the trick staggered collector for better ground clearance. It is hard to believe that this is a stock piece. The manifold also has brackets for adding the stock heat shields if you want to go stealth.

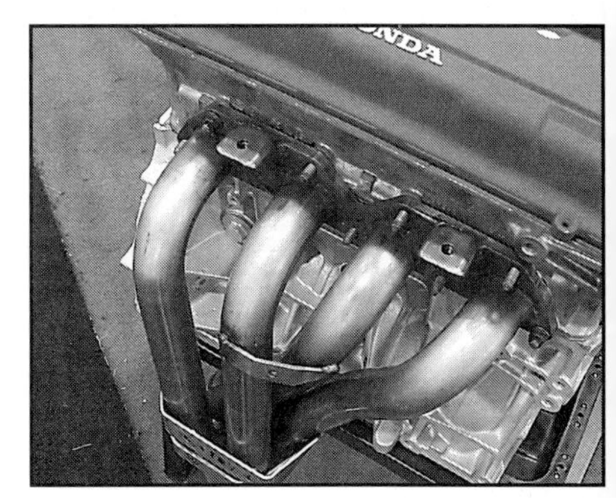

Here is the JDM ITR manifold mounted on a motor.

NITROUS OXIDE INJECTION 7

BASIC THEORY

Nitrous oxide injection or "the squeeze" as it is known on the streets, is probably the quickest and most cost effective way to get a boost in horsepower and torque easily, safely and cheaply. Although nitrous has fallen from favor as the primary power adder for Quick Class and higher competition, it is still a potent weapon for grudge and bracket racing as well as giving a street machine a fun burst of power available at your fingertips.

How does nitrous work? It is really quite simple; nitrous is a chemical supercharger. An engine creates power by harnessing the heat energy created by the combustion of fuel. Combustion is simply the rapid oxidation of fuel. In order to have the combustion of fuel needed to make power, you must have oxygen for it to burn. In an engine, the amount of oxygen available is the limiting factor to how much fuel you can burn and how much power you can make. Basically, nitrous oxide is an oxidizer, an oxygen-bearing chemical that has more easily available oxygen than atmospheric air. Adding nitrous is like injecting concentrated air and extra fuel into your motor. The principle is the same as with a turbo or supercharger, except that it is a chemical process as opposed to a mechanical one.

The Edelbrock/JG/NOS Nitrous system is an example of a maximum power nitrous system. There is a complete independent fuel system, which includes a separate pump and regulator to ensure an adequate supply of fuel to the nitrous unit. This system is also a direct port injection system, the best way to ensure equal nitrous and fuel distribution for high horsepower applications.

Air is 23.6 percent oxygen. That means of the total amount of air gulped by the motor with every cycle, 23.6 percent of this is going to be the oxygen needed for the power stroke's combustion event. The air is mixed with the proper amount of fuel and burned, producing power. The power is limited by the amount of air and fuel that the engine can take in with each cycle. To get more power, this amount can be increased mechanically by compressing the air and fuel mixture to force more of it into the cylinders by supercharging or turbocharging, or it can be increased chemically. Nitrous oxide injection, the chemical method of forced induction is one of the best ways to get more oxygen into the motor safely through the wonders of chemistry.

Nitrous oxide, or nitrous, is 36% oxygen by weight. This means that it has a lot more oxygen than regular air. Nitrous is also nearly twice as dense as air at room temperature, so pound for pound, it contains much more oxygen. Nitrous is also a cryogenic liquid, with a very high vapor pressure. This means that nitrous must be kept at pressures higher than 745 psi at 70 degrees F to remain a liquid. 745 psi is the vapor pressure of the nitrous.

Since the vapor pressure of nitrous is so high in relationship to the normal atmospheric pressure, the nitrous will

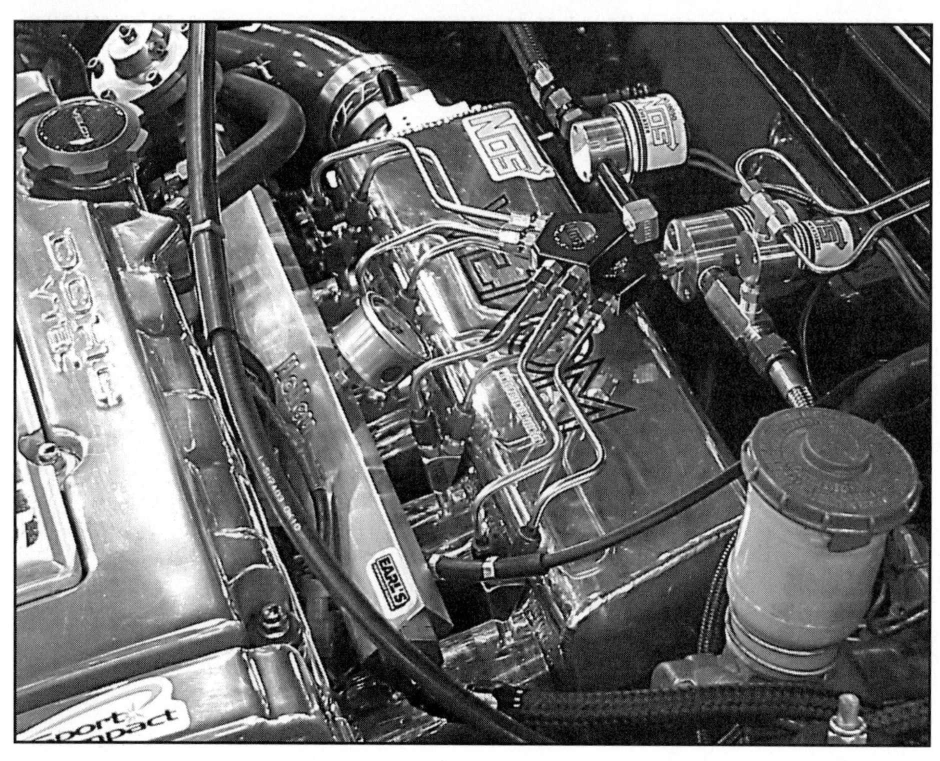

A menacing, full race direct port injection system by NOS graces this B16A-powered Civic.

undergo a very rapid expansion as it changes from liquid into gas when released into the atmosphere, where the normal pressure is only 14.7 psi. This is the same phenomena that occurs when boiling water but at a violently rapid rate. As the nitrous rapidly flashes into gas, a tremendous amount of heat is absorbed from the reaction, causing the intake charge to become supercooled to many degrees below freezing at the same time.

If you have ever done a long spray from a spray can or used a CO_2 fire extinguisher, you have probably noticed that the can or bottle started getting cold. This is the same phenomenon that is occurring with the released nitrous because the CO_2 in the fire extinguisher and the propane in the spray-can propellant have very similar physical properties to nitrous. In fact, nitrous is used as an aerosol-can propellant with certain food products, like canned whipped cream.

All of these properties are very useful when nitrous is injected into an engine. First, the nitrous itself has more oxygen than air, providing more oxygen for more power-producing combustion. Second, its oxygen density, especially when liquid, is many times higher than air. When liquid nitrous is introduced into the intake tract, it can add much more oxygen by volume than the air it displaces, even as it undergoes a phase change and rapid expansion from liquid to gas. Third, the near-cryogenic cooling effect of nitrous as it flashes into gas in the intake tract and the cylinder is very useful in cooling the intake charge, dropping the intake charge temp by at least 65-75 degrees. Cooling the intake charge reduces the propensity to detonate and lowers EGTs. Cooling also contributes to increasing the charge density.

Although nitrous is nonflammable under normal conditions, when it is heated past about 570 degrees in the cylinders of the engine as combustion starts, the nitrous decomposes into free oxygen and nitrogen, making the oxygen available to the combustion process. With the additional oxygen, you must have additional fuel, as the mixture will be much too lean and hot, burning to the point that severe engine damage will occur almost instantly.

A Bad Rap

This potential for sudden death is what has given nitrous oxide injection a bad name in the past. This bad rap is pretty much unwarranted as long as you do not stray from the nitrous kit manufacturer's jetting recommendations and take care to ensure that an adequate fuel supply is available under all anticipated operating conditions. The stoichemetric ratio for the complete combustion of nitrous and gasoline is approximately 9.7:1. This is quite a bit richer than the 14.7:1 of gasoline and air and is due to nitrous's higher oxygen content. For safety and a bit more power, a richer ratio of nitrous and fuel should be injected into the engine, approximately 8.5:1. If this ratio can be maintained under all conditions then nitrous is perfectly safe, even on stock engines when used in moderation.

Why Nitrous?

You might wonder why nitrous oxide is used as a combustion enhancer rather than some other chemical or pure oxygen. Well, one of the reasons is its ability to phase change rapidly and cause extreme intake charge cooling. The other main reason is that the nitrogen in the mixture acts like a combustion buffer,

The solenoids are really electrical, solenoid-activated high pressure gas valves. Here are NOS's fuel and nitrous solenoids up close.

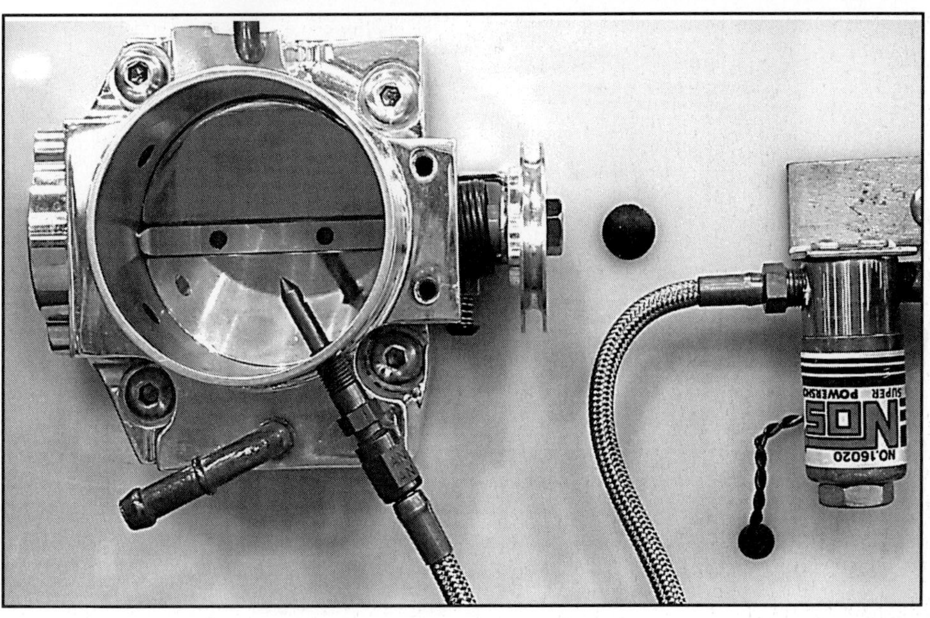

Here is a close up of the NOS single nozzle dry system.

ensuring that the combustion process does not go too fast in an uncontrolled manner, causing detonation. Pure oxygen is too aggressive, burning too fast and too hot. If you are familiar with oxyacetylene welding, then you know what happens when you add a lot of oxygen to the flame—you get a cutting torch. Imagine that inside your engine!

Other oxidizers, like nitromethane used by top fuel dragsters, are very tricky to work with. Nitromethane is very prone to pre-ignition and is difficult to tune correctly. Pre-ignition can be just as damaging to a motor as detonation. Pre-ignition can cause the same cylinder overpressure damage as detonation, but the cause of the overpressure is the mixture spontaneously igniting at an inopportune time rather than burning unevenly, as in detonation. Witness the many explosive demonstrations at any Top Fuel drag race. Even many of the top pros have trouble getting everything just right!

When you add it all up, nitrous when properly set up, is pretty safe as a power adder as long as a few precautions are taken. With nitrous on a 4-cylinder Honda engine, it is possible to get a reasonable safe power gain of up to 75 hp on a stock motor and pump gas. On a motor specifically built for nitrous, it is possible to get gains of over 250 hp on a 4 cylinder motor. Although this is not as much as what a full race turbo system can produce, there is still nothing that can give a bigger bang for the buck.

THE NITROUS SYSTEM

Nitrous systems are very simple. A typical basic nitrous system consists of a high pressure bottle of about 10 lbs. capacity or greater, high pressure braided steel nitrous lines, high pressure solenoid valves to control the flow of fuel and nitrous, and injection nozzles with adjustable, changeable orifice sizes or jets. These jets, much like old-school carburetor jets, allow tuning by controlling the volume of nitrous and fuel injected into the engine. Bigger jets have a bigger orifice and more flow, smaller jets have a smaller orifice and less flow.

As the power level and complexity of the system increase, bigger nozzles and sometimes more nozzles are used to deliver more fuel and nitrous. The fuel system must be modified to ensure an ample supply of fuel and peripherals like bottle heaters and progressive solenoid controllers come into play.

The higher cylinder pressures of big nitrous systems also require more powerful ignition systems; ignition timing controls and even tougher bottom end components such as forged pistons and beefed up connecting rods. In the case of the Honda, open deck motors can benefit from block savers, thicker sleeves and even a closed deck conversion to help handle the extra pressure, just the same sort of preparation as a high boost turbocharger requires.

MILD STREET SETUPS

For a completely stock Honda motor running on 92 octane pump fuel, it is not good to exceed much more than 50% of the motor's rated power or 75 hp whatever comes first.

The ZEX nitrous system is easy to install, self-regulating and nearly foolproof. It is a very good system if you desire less than 100 hp. The ZEX system is a good example of a single point dry system. All of the metering components for the system are located in the purple box.

For frequent use or for use longer than a 1/4-mile or a 15-second blast, the nitrous system should be limited to 50 hp. With any nitrous system use, especial one intended for mild street use, the total single use time should be kept to blasts of 15 seconds or less for safety. Although it does not sound like a lot, it is pretty hard to keep your foot to the floor for more than 15 seconds anywhere, even on a racetrack. As the jump in cylinder pressure can cause detonation at low engine speeds, the nitrous system should not be activated below 3500 rpm. Most mild street set-ups are single point injection.

Single-Point Injection Systems

A Honda motor can take up to 75 more hp on a simple single-point injection street-type nitrous kit. This is pretty close to the limit of the connecting rods of the D-series motors. For frequent use on strictly pump fuel, 50 hp is probably more appropriate and goof-proof for most people. Single-point systems can be divided into two categories—wet or dry systems.

Dry systems have a single nozzle and one nitrous solenoid. The nitrous is introduced from the solenoid through this single nozzle. To handle the needed additional fuel, a special pressure regulator that increases the fuel pressure to the injectors while the nitrous solenoid is activated provides enrichment.

The ZEX System—The ZEX nitrous system is an easy-to-set-up system that contains the solenoid, the nitrous jets and the nitrous pressure referenced fuel pressure regulator in a self-contained box. To get additional fuel enrichment, the return line to the gas tank is run through the ZEX pressure regulator located in the purple box. When the nitrous system is engaged, nitrous pressure is routed to the backside of the ZEX regulator's diaphragm. This causes it to clamp down on the return fuel line, returning fuel to the gas tank, causing the fuel

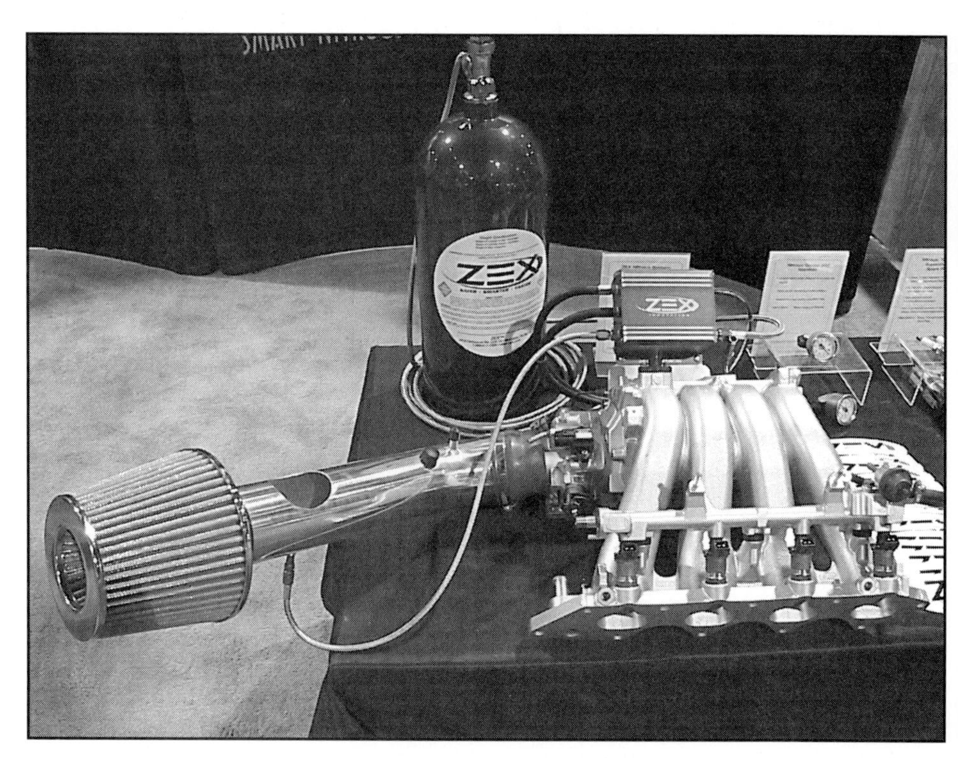

In this cutaway air intake you can see the injection nozzle of the ZEX system. The ZEX system uses a single nozzle to spray nitrous only into the intake stream. The ZEX nozzle should be mounted at least 8" before the throttle body for the best mixing of nitrous and intake air.

pressure to rise. This provides the additional fuel enrichment through the engine's stock injectors. The cool thing about the ZEX system is that since it references the pressure of the nitrous being injected, it can somewhat compensate fuel flow to the bottle pressure as it changes during a run. It self-adjusts the injectors' fuel flow to match the nitrous flow by regulating the fuel pressure. The ZEX system is adjusted by changing the nitrous metering jet in the nozzle and the nitrous pressure bleeder jet in the regulator control box. The ZEX system has other nifty accessories available such as a remote bottle opener and a bottle heater

NOS—Nitrous Oxide Systems or NOS, a pioneer in the field of nitrous oxide, also has Honda dry systems available. The NOS system also uses nitrous line bleed pressure to clamp on the opposite side of the fuel pressure regulator, pinching off the fuel return line to increase fuel pressure during system operation. It is a simple, effective low-priced system. JG/Edelbrock also has a dry-type single nozzle nitrous system with very similar capabilities as the NOS unit.

Venom—Venom has an advanced computer controlled dry system that electronically controls the injectors for additional fuel. The Venom VCN-2000 also electronically controls the nitrous flow by modulating the nitrous solenoid to ensure a stable mixture so no metering jets are needed. A neat feature of the VCN-2000 system is that it uses feedback from its own O2 sensor to continually adjust the nitrous/fuel mixture ratio. The throttle position for injection and the injection amount taper is all electronically adjustable. Although much more expensive than the other single-point injection dry systems, the VCN-2000 system is highly flexible and versatile. Due to the Venom system's sophisticated electronic controls and its ability to control larger than stock injectors, it can make from 25 to 180 more hp.

A close-up of a NOS dry nozzle shows the tip designed for a fan-shaped spray pattern. The fan spray allows the NOS nozzle to be mounted right on the throttle body and still have good mixing with the intake air.

The advantage of the dry system is that you have even distribution of fuel, as modern Honda manifolds are designed to flow a dry gas, not liquid fuel. Adding the fuel through the injectors ensures that even distribution of fuel will occur. By not needing a fuel solenoid and line, a dry system can also be cheaper and is easier to install.

The disadvantage of most dry systems is that you can only squeeze so much fuel flow out of the stock injectors and fuel pump. This limits dry system power to a safe 50 or so hp, although some companies claim that up to 75 safe hp is possible. As previously stated, for frequent nitrous use on a totally stock engine, it is probably a good idea to run no more than 50 hp for long-term reliability. A 50 hp shot is still capable of shaving at least 1.5 seconds of the quarter mile time of a small Honda engine.

Foggers—The other type of single nozzle setup is one using a fogging or mixing nozzle mounted several inches before the throttle body. The fogging nozzle mixes the nitrous and fuel together before spraying it into the intake tract. This is known as a "wet" system. This is important in a single point injection system, as mixture distribution can be a problem. As we mentioned before, the manifolds on modern Hondas were designed to flow dry air, not a mixture of air nitrous and fuel. If the fuel and nitrous can be more of a homogeneous cloud to suck into the engine rather than separate streams of either, the better the distribution of the mixture. This is why it is better to put the fogging nozzle back from the throttle body by about six inches instead of right on the throttle body like some kit makers recommend.

Since the fogging nozzle has its own built-in fuel supply by having an additional fuel solenoid and line, the flow limit of the stock injectors is not a factor. However, to be safe, a single fogging nozzle system should not be run at more than 75 hp or unequal distribution of fuel may lead to problems due to unequal wet flow and fuel condensing on the walls of the manifold. Usually, cylinders one will be lean and burn plugs while the end cylinders will receive a progressively richer mixture, if jets much larger than this are used. Nevertheless, people have run up to 100 hp with no problems on systems like this with proper tuning. Like we mentioned before, it is best to stick with the jetting recommendations of the kit's manufacturer to be safe. Nitrous Oxide Systems and Nitrous Express make proven wet, single mixing nozzle nitrous systems. Nitrous Express has a unique nozzle with 4 ports that allows a two-stage system with a single fogger nozzle.

There is a debate over which type of system, wet or dry, works better with industry experts being divided over which system is best. In my opinion, for systems of up to 50–60 hp, the dry systems are usually more reliable due to the even distribution of fuel. However, a dry system starts to get iffy beyond this due to injector capacity, as there is a limit to how much fuel can be reliably squeezed from the stock injectors by just increasing the fuel pressure. For slightly more power than that, the wet systems can work well but caution must be used with this route. Either way, if you pick a proven nitrous kit maker and follow their jetting recommendations, ensure that your fuel system has a good fuel pump and a clean fuel filter, you should be safe.

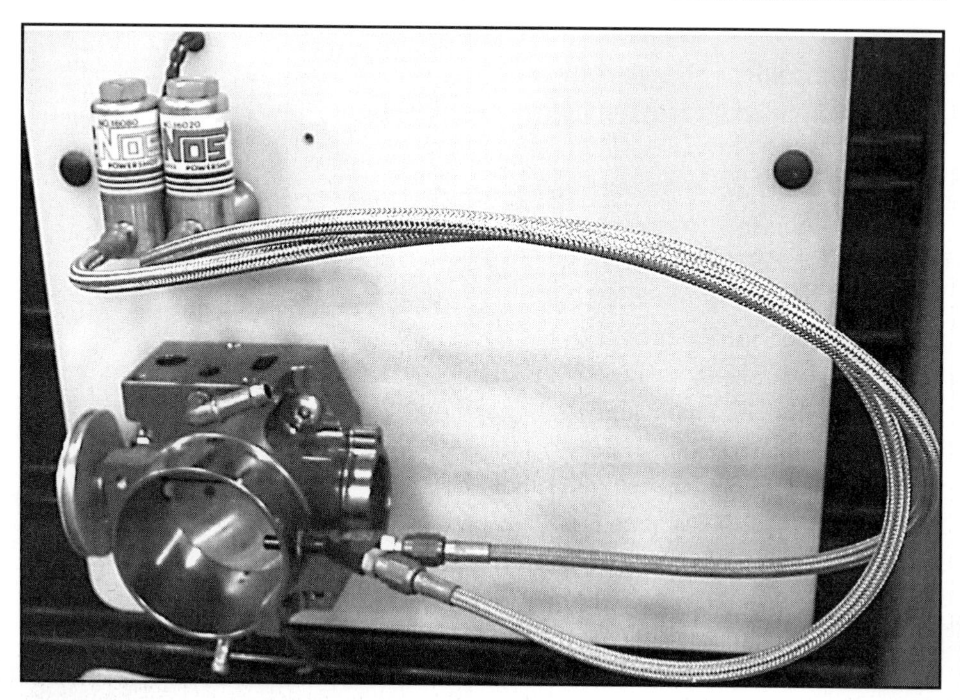

The NOS Fogger nozzle mixes the nitrous and fuel together before injecting it into the engine in a finely blended mist of fuel and nitrous. Fogger nozzles offer superior mixing and safety to the old-school way of having separate fuel and nitrous injection nozzles. This application is a single nozzle fogger. Foggers are known as "wet" systems.

Fuel—For these smaller streetable systems, nothing more is needed other than 92 octane fuel, one heat range colder spark plugs. For kits over 50 hp, you might need to retard timing 2–4 degrees. These kits are a fun and cheap performance benefit giving a huge bang for the buck.

Slightly Modified Engines—If your engine is mildly built with perhaps JDM R pistons, or if you have a B18C5 Type R Integra motor or a JDM B16A Civic Type R motor with high 11:1 compression, then a 50 hp shot kit is the safe limit for that high compression with pump gas. Even then, caution should be used, with perhaps a two-heat range colder spark plug and the timing retarded 4–6 degrees when the nitrous is used.

RACE SYSTEMS, DIRECT PORT INJECTION, TWO-STAGE SYSTEMS

For systems producing more than 75 hp, direct port injection is recommended. Direct port injection uses a fogging or mixing type nozzle for each cylinder's intake port. Instead of being injected upstream of the throttle body, a direct-port injection system introduces the nitrous and fuel right into the runner of the intake manifold, as close to the head's intake ports as possible. This ensures even distribution of fuel and nitrous when going to bigger systems with additional power. A fuel and a nitrous solenoid are used, directing fuel into separate distribution blocks. From the blocks, the fuel and nitrous are plumbed to the injector nozzles mounted in the manifold.

Spray Bar Plates

Nitrous and fuel spray bar plates can also be used for high horsepower

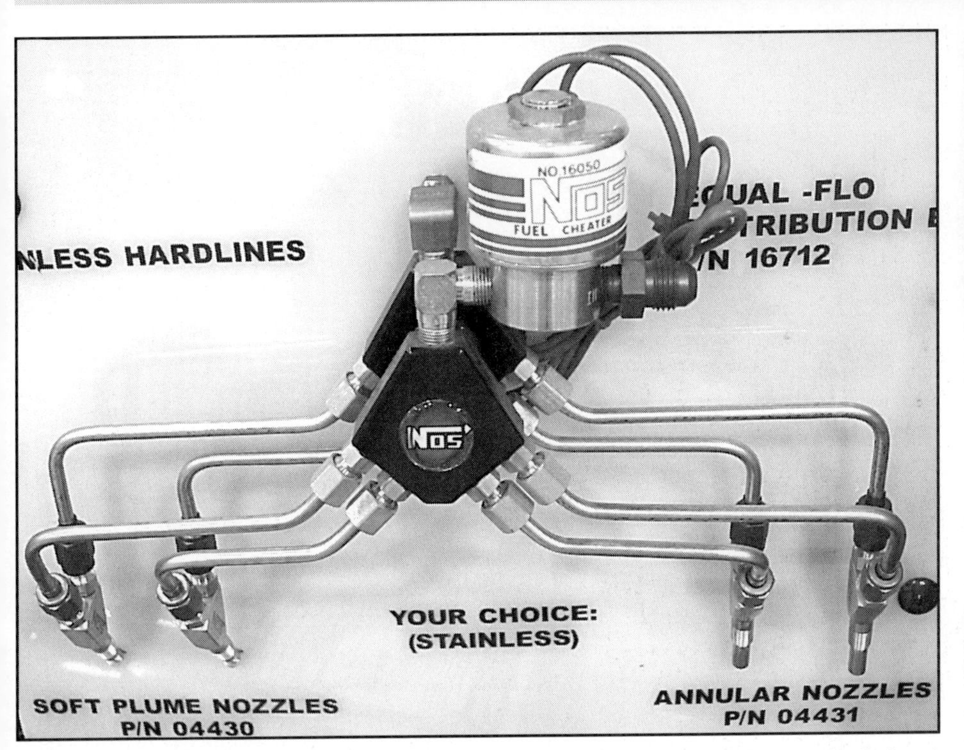

The working parts of the direct port injection system are shown here off the car: the fuel and nitrous solenoids, hard lines, fogger nozzles and distribution blocks.

Here is a full race direct port injection nitrous system installed on a STR box plenum intake manifold. The blue and silver cylinders with the wires attached to them are the fuel and nitrous solenoids. The steel hardlines go from the solenoids to the distribution blocks, then to the fogger nozzles mounted into the manifold. The fogger nozzles contain the metering jets.

systems. In a plate system, a specially machined plate is placed between the halves of the manifold. This plate contains spray bars for fuel and nitrous with metering orifices drilled into the spray bars to control the

A close-up of the NOS fogger nozzle. The nitrous and fuel metering jets are under the colored fittings where the hard lines attach to the fogger nozzle body.

The business end of the fogger nozzles are shown here when the end plate of the manifold plenum is removed. As you can see, the nitrous/gas mixture gets a straight shot into the cylinders with no chance of unequal distribution possible.

volume of both the fuel and nitrous. Because they do not offer the superior mixing of fuel and nitrous that fogger nozzles offer, plates are falling out of popularity. Plates will only work on Honda motors with two-piece manifolds like the B18C anyway.

Two-Stage Systems

Two-stage systems are just like they sound, they run two stages, usually a smaller single nozzle system before the throttle body for the first stage and a bigger direct port injection system for the second stage. The small stage helps prevent from overpowering the tires out of the hole and the second

Here is a plate on a B18 system. In this custom system, a single nozzle in the throttle body introduces the nitrous, while the plate in the manifold runners adds the fuel. This ensures safe fuel distribution.

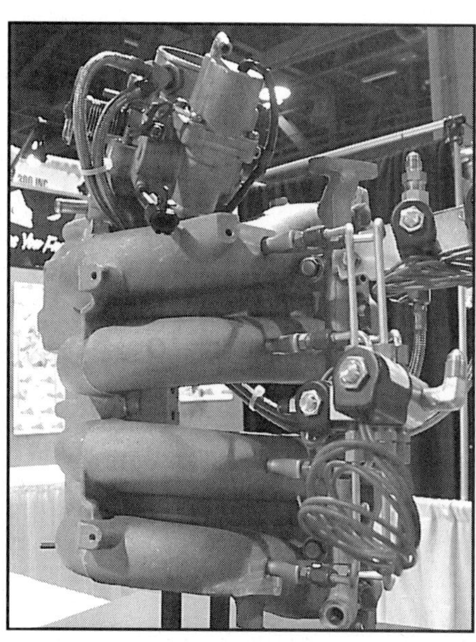

This is a Nitrous Works two-stage system. The single fogger in the throttle body is a smaller system of about 50-75 hp. This system is engaged out of the hole to keep from blowing away the tires. Once the car is well under way, the second stage, the direct port system, is activated. The direct port system usually adds 100-150 hp. The main advantage of a two-stage system is that it won't smoke or blow the tires as easily off the line.

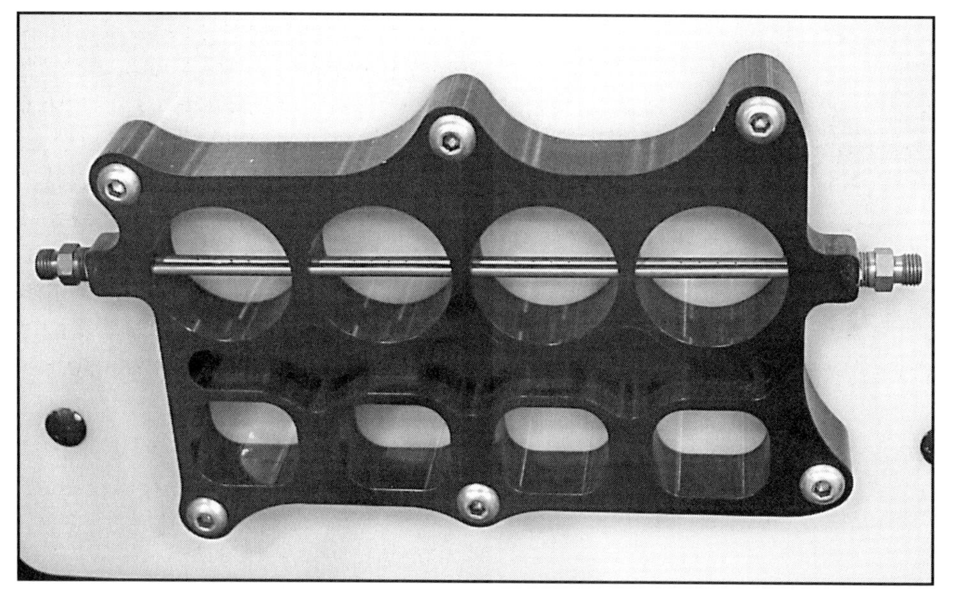

Here is a close-up of the NOS B18 spray bar plate. Although it works very well to distribute fuel and nitrous evenly, NOS's direct port fogger nozzle system will work even better. The main advantage to the plate is that it is quick and easy to install. No drilling and tapping of the manifold, no bending and fabricating hard lines.

stage is kicked in after the car is moving, at higher rpm, preferably after the torque peak when the engine's natural cylinder pressure is dropping and the motor can tolerate more nitrous. The advent of the electronic progressive nitrous solenoid controller is rapidly making the two-stage system obsolete.

Engine Modifications

With direct port injection, it will probably be necessary to upgrade the fuel pump for a higher volume. Forged pistons will be needed. If stock or JDM cast pistons are used, then extreme care will have to be taken to avoid detonation. A higher-powered ignition becomes needed on Hondas at above the 120 hp or so point and having a retard option on the ignition system is handy to retard the spark under nitrous operation. At much past the 100 hp mark, it is probably a good idea to run race gas, at least some of the lower octane unleaded race gasses. Spark plugs that are 1–2 heat range colder are needed to prevent them from melting and becoming a point of detonation. It might also be necessary to close the plug gap down to as little as 0.020" to prevent misfire. It is also advisable to pull 4–8 degrees of timing out during

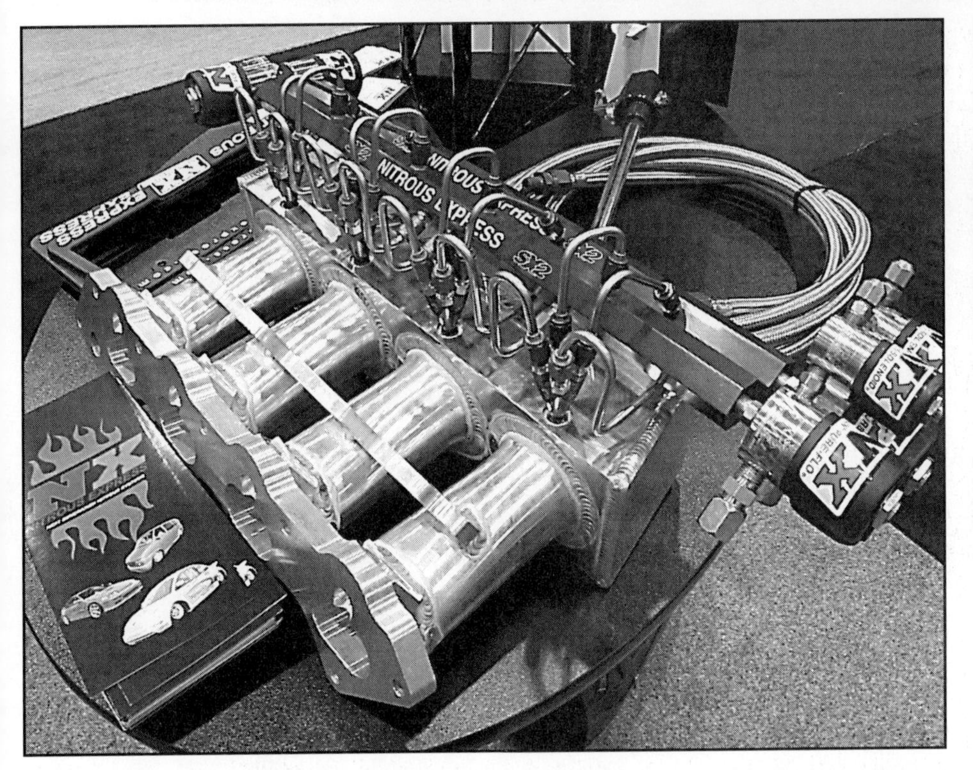

The monstrous Nitrous Express two-stage system installed on a B18 manifold uses the NX unique 4-jet fogger system. Each nozzle can handle two nitrous and two fuel jets at once. This makes it good for two-stage use. The NX fogger can also outflow any other nozzle by twofold. This nozzle can inject all the nitrous an engine can handle.

A close up of the NX nozzle shows the four input ports. The metering jets are located under the red and blue fittings, red for fuel and blue for nitrous.

nitrous operation at this power level.

When going much past the 300 total hp point, or a 150 hp shot, it is a good idea to do the same block modifications that you would do on a turbo motor of about the same power level. This includes such things as racing rods, closed deck modifications, block sleeves and mandatory use of forged pistons. A good ignition system, an ignition retard system and race gas becomes mandatory at this point. When pushing 400 hp or a 250 hp shot, it is a good idea to reduce the compression ratio to about 8.5:1, just like you would do with a turbo system. Remember, a 250-hp unit on a 4-cylinder is like a 500 hp super duper shot on a big V-8 when it comes to cylinder pressure. When squeezing this hard, the use of a high specific gravity racing fuel like VP C16 is also advisable. Many racing fuel companies have specific fuels for big nitrous units or a lot of boost and it should be used exclusively at this point. When tuning for a particular fuel, it is best to stick with it as the fuel specific gravity can make a big difference when tuning for best mixture. When a shot this big is being run, it will also be necessary to pull 6–10 degrees of timing out when the nitrous is being run.

System Limits

A system with 420 total hp is close to the practical limit of a Honda nitrous system at this point for two reasons.

Unlike a big V-8 Pro-Mod car, a small 4-cylinder has trouble ingesting enough nitrous to make more power than this at this power level. Theoretically, when calculating the power gain based on a chemically correct mixture of nitrous and the engine's swallowing capacity, it is only possible to make about 250% more power than stock. The actual possible limit is slightly higher than this because of the difficult-to-predict effects of intake charge cooling and percent of liquid nitrous ingestion into the cylinders. However, it is easy to see that all-out racing turbochargers are capable of a greater power boost than this.

When this much nitrous is being injected, the nitrous flow volume is so great that the bottle pressure fluctuates and drops considerably during the course of a 1/4-mile run, even when starting with a full bottle. This makes it difficult to jet the system so it runs consistently and safely over the course of a run. A full bottle at 950 psi might only be at 700 psi or less when the pass is over. Jetting the system so it is safe at the line with a full bottle means that the engine will be running rich with less power at the end of the quarter. If you jet for peak power at the top end, you risk being dangerously lean out of the hole. At this power level the bottle ends up being more than half empty at the end of a run so that means having a lot of bottles handy for an event or owning your own fill station.

This is probably why nitrous equipped cars have fallen out of favor in the Quick Class and faster classes. The racers have found the limits of nitrous with a small four-cylinder engine to be in the high 10-second, low 11-second range, not fast enough to ensure making the qualification cut in the Quick Class today.

TURBOS AND NITROUS

Nitrous works very well with turbochargers. For mild use, nitrous can be used to help spool a big turbo. With its instant boost of power and greatly increased exhaust gas volume, even a small nitrous system can rapidly spool the biggest turbo, nearly eliminating lag. This kind of system uses a pressure switch to automatically shut off the nitrous when a preset boost pressure is reached. An advantage of this sort of system is that a very large turbine wheel and housing can be used to greatly lower the turbo's backpressure on the engine. This can greatly increase power, but these bigger parts usually greatly increase lag. A nitrous spool assist can compensate for these freer-flowing turbo parts.

The hardcore will want to squeeze all the way through boost. This is advisable only for serious racers as it is quite punishing to the engine. Usually with a big, free-flowing turbo, very good, turbo-specific racing fuel, and fine tuning, a 50 hp shot is no problem at up to 25 psi or so of boost. A larger turbine housing and possibly a larger turbine wheel is recommended if you are planning to do this. The hardcore, after every last bit of power, have done up to 120 hp shots with up to 30 psi of boost. Expect to hurt some stuff occasionally at this level even if you do everything right! Hondas need every modification to the bottom end to survive this power, especially the closed deck conversions. Benson's full metal jacket or a deck plate or the cylinder walls will crack with this sort of load.

An interesting note is that not too many of the top Quick Class cars need both turbo and nitrous to run in the nines, so if you feel the need to squeeze the nitrous on top of a lot of boost, consider this as it can get expensive. With today's advanced turbo technology, the use of nitrous with a properly designed turbo system is not really needed to be competitive.

NITROUS ACCESSORIES

There are a few things that can make nitrous use easier, safer and better performing. Some of them work so well that they are practically mandatory for proper nitrous system operation.

Bottle Heater

A bottle heater is a thermostatically controlled electrical heating blanket that helps keep the nitrous bottle at a constant temperature to help maintain a more equal pressure throughout the charge of nitrous. As the nitrous is discharged, the bottle is severely chilled and the pressure drops. A bottle heater can get the nitrous bottle back up to pressure, ready for another full powered run in just a few minutes. Most nitrous systems are jetted to run their best at 900–950 psi and a bottle heater is the most important tool to help maintain that pressure. With a bottle heater, it is important to hook up the thermostat safety shutoff switches and to ensure that the safety blow off valves of the bottle are in proper working condition. Never hook the bottle heater directly through the battery by tapping into non-key controlled power. Recently, a spectacular explosion of a nitrous bottle caused by improperly wired bottle heater overheating the bottle, destroyed a car and partially the garage of an unsuspecting nitrous user.

ZEX has a very nice bottle heater where the thermostat, relays and heating element are contained in a single unit.

Bottle and Fuel Pressure Gauges

Pressure gauges allow you to monitor the system pressure. It is very important for tuning and for consistent runs. If you always start a run or tuning session at the same bottle pressure, you can have a better indication that your motor is getting pretty close to consistent amounts of nitrous from run to run. Pressure gauges with a bottle heater are important tools for the serious-minded performance enthusiast. A fuel pressure gauge is mandatory when attempting to tune a system

This NOS direct port injection system has a fuel pressure gauge built right into the fuel rail so the fuel pressure closest to the injection point can be observed. When tuning big units like this with an adjustable regulator, accurate measurement of the fuel pressure is critical for safe tuning.

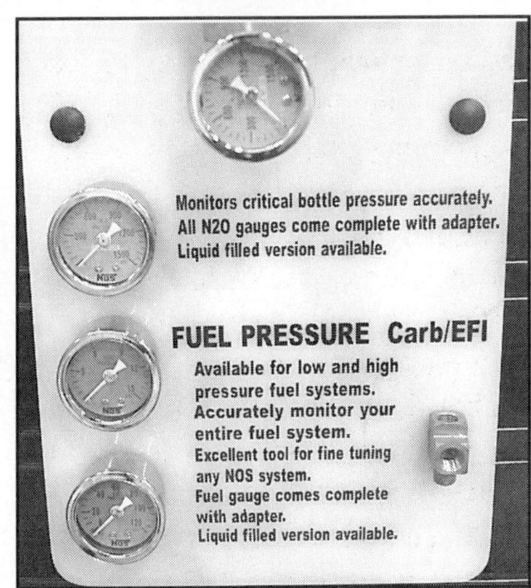

NOS among others, sells an assortment of fuel and nitrous pressure gauges. Monitoring these pressures is critical for tuning and for when consistency is important, like when bracket racing.

Ignition Retard Control

As it is a good idea to retard the ignition (the amount depends on the size of the nitrous system), an ignition timing control, like those sold by MSD, is extremely useful. This device allows you to automatically retard the spark when the nitrous system is activated. This way you don't have to drive around with soggy retarded timing when not running the nitrous.

Progressive Nitrous Controller

A progressive nitrous controller cycles the solenoids to a duty cycle, much like a fuel injector, to precisely control the onset of the nitrous's power increase. The onset of the power can be automatically adjusted with one of these controllers. This helps prevent using too much nitrous at too low of an rpm, possibly damaging the engine. It also helps traction out of the hole. Engines can take more nitrous at high rpm after the torque peak as the volumetric efficiency and the cylinder pressure start to fall off. As the cylinder pressure decreases, the engine can tolerate more nitrous safely without the risk of engine damage. Progressive controllers have just about rendered the old-school two-stage nitrous system obsolete. NOS

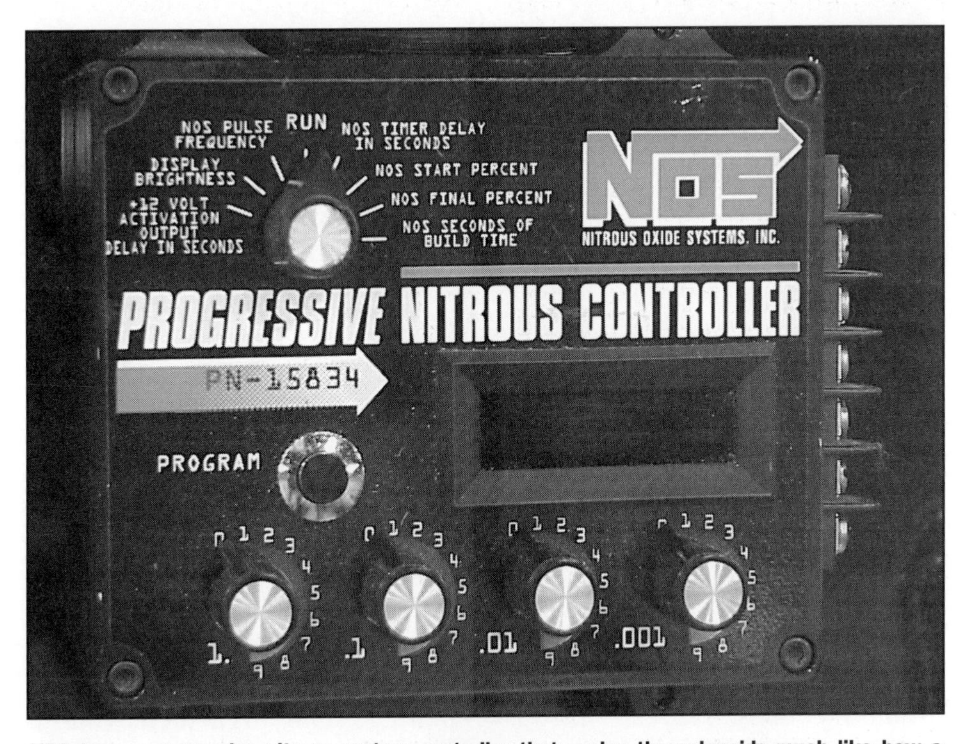

NOS has a progressive nitrous system controller that cycles the solenoids much like how a car's ECU controls fuel injector pulse width to meter the nitrous into the engine gradually. This prevents a sudden spike of power from blowing the tires away, losing traction. NOS solenoid will wear the valve seat quickly with one of these controllers but at least the solenoids are quickly and cheaply rebuilt with an easy-to-use kit.

The NX controller works much like the NOS system except that NX solenoids have a much more durable valve seat and are better suited to long-term rapid cycling by the control unit with no damage.

The ZEX remote bottle opener uses a gear motor to operate the bottle's main valve.

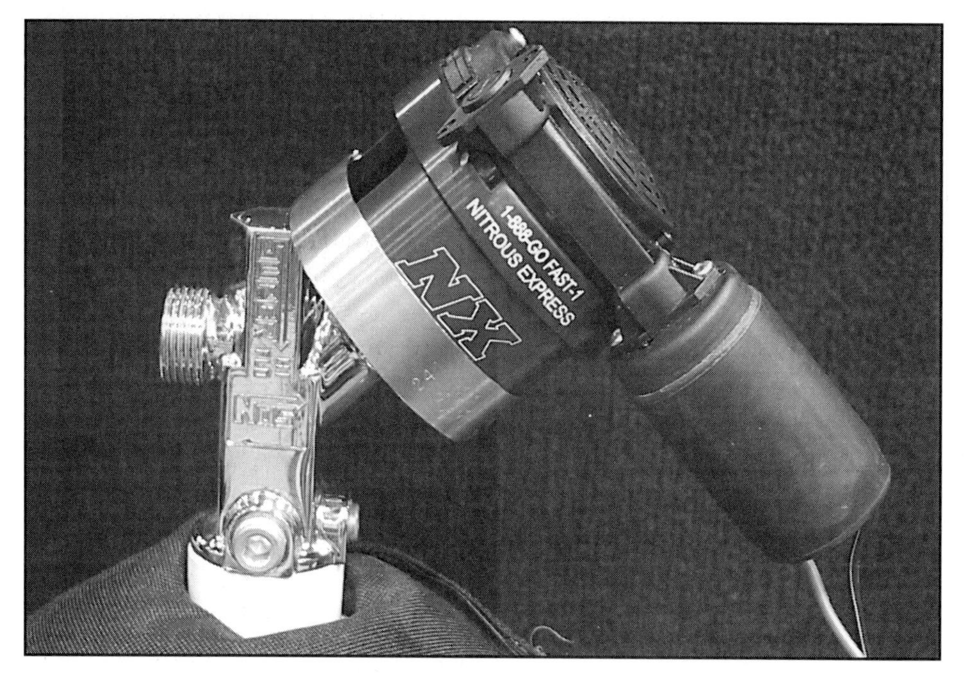

Nitrous Express also uses a gear motor to open and close the valve. The NX opener also works on bottles from NOS.

sells three types of progressive controllers: A time delay unit that gradually increases the nitrous volume over time which is adjustable; a programmable unit that adjusts the percentage of nitrous injected in stages over time; and a throttle position sensitive controller that adds nitrous in proportion to the throttle opening. The high-tech Venom system has a built-in progressive control as a standard feature.

The only disadvantage to a progressive controller is that they tend to rapidly wear the Teflon valve seat of the nitrous solenoid. This results in greatly reduced performance as a worn dimpled solenoid seat greatly blocks the flow of the solenoid. Fortunately solenoids can usually be cheaply and easily rebuilt with simple hand tools by anyone with enough skill to install the system in the first place.

Remote Bottle Opener

When leaving a nitrous system unused for more than a few hours, it is important to close the shutoff valve on the bottle. This is because the nitrous solenoids are not perfect seals and the bottle will empty from seepage in a few days with just the solenoid valve holding the nitrous pressure back. The heavier-than-air nitrous can accumulate in the intake manifold while the engine is sitting overnight and when an attempt is made to start it the next day, the nitrous-filled manifold can explode with nasty results.

However when keeping the bottle closed, it always seems that when it is needed, the closed valve is in the

A two-stage system needs four solenoids, two fuel and two nitrous to work. Each pair of solenoids are triggered separately. On this NX solenoid pack, the small solenoids are for purging air and gaseous nitrous from a system before a run. This ensures a solid hit of pure liquid nitrous right out of the hole to prevent bogging.

trunk and you are in the driver's seat of a moving car. The automatic remote bottle opener gets around this problem. The NOS remote bottle opener is a heavy-duty solenoid valve that mounts on the bottle. This solenoid has enough closing pressure to be leak free and allows you to leave the bottle valve open all the time, with nitrous available at a push of a button. ZEX and Nitrous Express make remote bottle openers that use a gear-driven electric motor to work the bottle valve through a remote switch. The Nitrous Express opener will work on NOS bottles, which is a big plus. This is a more secure, safer way to do a remote bottle opener. A remote bottle opener is the ideal accessory to swat the pesky muscle car that may be hassling you, not that we condone street racing or anything like that, of course.

High Volume/Pressure Fuel Pump

Big nitrous systems require a big supply of fuel to run to their full potential and to run safely. There are a large number of fuel pump makers on the market to choose from. A big

Here is a more simple NOS purge system; you can see the small purge solenoid, which is installed right before the system's main solenoid.

nitrous system needs a pump that can flow about 250 liters per minute at 40 or so psi. The fuel lines, fittings and filters must also be up to the task of moving a lot of fuel. See Chapter 9 on turbochargers to calculate what size of fuel pump you will need for the anticipated power you want to produce.

Some tuners have an independent fuel system so the fuel injected with the nitrous can come from a separate tank full of super high-octane race gas. That way the car can be run cheaply on regular pump fuel, but the nitrous system can be fed racing fuel when it is needed.

Purge Systems

The nitrous in the line between the nitrous bottle and the solenoids tends to vaporize if the system is not used for several minutes. This nitrous vapor is much less dense than liquid nitrous. When engaging the nitrous system, it takes a fraction of a second for this nitrous vapor to be expelled and the liquid nitrous to reach the nozzles. This can cause a bog because there will be too much fuel injected for a fraction of a second. The bigger the nitrous unit, the more noticeable this can be.

To prevent the bog, a purge system can be installed. A purge system is a small solenoid connected to a vent tube installed right before the main nitrous solenoid. Usually the vent tube is installed to discharge where the driver can see it. The purge solenoid vents the nitrous vapor so that pure liquid nitrous will hit when the system is engaged. The vent solenoid is opened by a driver controlled push button until the nitrous vapor can be seen by the driver. This means that the system is fully purged and is ready to go.

Intercooler Cooler

Nitrous Express has a kit to spray cold nitrous on the intercooler to help keep it extra cold during a run. Since it is illegal to drip water on the track, using dry nitrous can help prevent the disqualification that a water sprayer might cause at a race. Don't ask us how we know that.

TUNING YOUR NITROUS SYSTEM

First off, if you have a small street-type nitrous unit, stop right here and don't proceed any further. You don't need every last bit of power and you should stick to the manufacturer's kit recommendations. If you are a real racer and need a little more power out of your big, full-race unit, proceed carefully. Tuning a full race nitrous system requires a lot of experience in reading plugs, some restraint and a lot of common sense. It is not recommended that you deviate from the manufacturer's recommendations for jetting, timing and fuel pressure unless you have a good handle on tuning and some experience with tuning nitrous, preferably under the guidance of another experienced person or on someone else's motor! Unfortunately, the best nitrous tuning experience is sometimes best gained by breaking some stuff!

Usually the factory jetting of most nitrous systems, even race systems, is on the very conservative side. This is best for the nitrous companies but possibly not for the serious racer. With careful tuning it is possible to easily extract an additional 5–20 hp from most factory kits.

Tuning should be done with good gas. A high specific gravity fuel designed specifically for nitrous and boosted applications will work best, like VP's C16 fuel blend. Because of nitrous's high oxygen density, more power can be gained by leaning out the nitrous and fuel mixture over advancing the timing. In fact, it may sometimes be better to lean out the mixture and retard the timing more, rather than advancing the timing.

Nitrous Express offers a unique intercooler sprayer that chills the intercooler with near cryogenic liquid nitrous. This really drops the charge air temperatures for your turbo. Since dripping water on the track can get you disqualified, this sort of spray system is better than the more common intercooler water spray devices.

Leaning the Mixture

The most power can be gained by leaning the mixture out with an adjustable fuel pressure regulator. Be sure your bottle is reasonably full and you always start a tuning run at the exact same bottle pressure with every tuning pass. Nitrous and fuel pressure gauges are mandatory and a good nitrous bottle heater is very useful. Reduce the fuel pressure by a couple of psi every run, and check the plugs after every run for any signs of detonation. If any detonation is heard, even the slightest bit, the throttle must be lifted immediately. An engine on a big nitrous unit can be destroyed in about two seconds of hard detonation. If you have reduced fuel pressure by 10% and there is no sign of detonation and the engine is running stronger, you may want to return to your base fuel pressure and go to one size smaller of a fuel jet. Do not reduce your nitrous system fuel pressure by more than this.

Bottle Level

Remember, do not start leaning out the system on a nearly empty bottle. Many a racer has gotten engrossed in tuning, not paying attention to the bottle fill level. Thinking he had hit the perfect tuned setup with a nearly empty bottle, he blows his motor sky high as soon as he installed a nice full bottle, leaning out the system to the max! When tuning, the bottle should be changed before the half full point, waste or not. You must always start a tuning pass at exactly the same bottle pressure also.

If you have gone one size smaller in the fuel jet and the engine still wants less fuel, you might want to try adding two degrees of timing up to the base stock timing. Be very careful, as a big racing nitrous system will usually feel better and better until it blows up! Be sure to read the bit on checking plugs in our basics section in the beginning of the book.

To do finer tuning than this, you might want to hire the help of a very experienced pro, but even pros can mess up at this point, especially with larger systems. As we stated before, tuning beyond the manufacturer's recommendation should only be considered if you are a pro racer that needs every last bit of power and are willing to take a bit of a risk to get it.

This is about all of the ins and outs of nitrous use. In small systems, it is a safe power adder that gives a good bang for the buck, which can really bring a car to life. Bigger systems get bigger results but can be challenging to run. Either way, nitrous is a powerful, relatively inexpensive performance tool that delivers the most bang for the buck!

SUPERCHARGING THE HONDA ENGINE 8

A supercharger is a device that uses a compressor to increase the density of the intake air charge. By increasing the density of the intake charge by pressurizing it, the oxygen content of the intake mixture is dramatically increased. As a result, more air is induced into the engine's cylinders. When more fuel is added to this denser mixture, it has much more explosive power to be released during the combustion event. Simply put, a supercharger is like stuffing a bigger engine into the same space occupied by your smaller engine.

Although superchargers are not very successful in heads-up import drag racing because they lack the raw power of turbochargers, they are an excellent choice for the street. A supercharger can usually be bolted on in a weekend and can add as much as 40 to 100 horsepower. About the only drawback is that they are relatively expensive.

The Vortech centrifugal supercharger kit for the Civic Si is an impressive piece, especially this one with the optional air-to-water intercooler. This kit cranks out an impressive 277 hp to the wheels, and it is smog legal too. Check out the cool carbon fiber work to the engine.

Most of the current supercharger kits are either CARB approved or are in the process of obtaining CARB approval, so there is less hassle when getting a smog check, especially if it includes a visual inspection. Even if you only have to pass a tailpipe smog check, a supercharger affects emissions less than other modifications. This is a significant advantage. A supercharger is probably the most powerful 50-state-legal power adder that you can get.

Another advantage that superchargers have is that since they are directly connected to the crankshaft, they give predictable, repeatable throttle response. This is a supercharger's primary advantage over turbochargers.

An interesting technical advantage to a supercharger is that on low boost, street-type applications, it is possible for a supercharger to out-power a small turbocharger. This is because a supercharger, despite its 10–20 hp drag on the crankshaft, does not have the backpressure-inducing turbine that a turbocharger has in the exhaust stream. As a result, because of its much lower exhaust backpressure, a supercharger has better volumetric efficiency than a turbocharger, especially a small quick spooling turbo like those often found in low-boost street kits. This often enables a supercharger to make more power than a turbocharger under the low 6–8 psi boost pressure often found on street-only systems. The turbocharger's big advantage does not come until larger, less restrictive, turbochargers are fitted and higher boost levels are run in the 12 psi and up range.

TYPES OF SUPERCHARGERS

There are basically two types of superchargers: positive displacement superchargers like the Roots supercharger and centrifugal superchargers like the well-regarded Vortech and the Powerdyne. Both types of superchargers have totally different power characteristics from each other. It is important to understand the characteristics for you to choose which type best suits your needs.

Roots Supercharger

Let's first discuss the Roots positive displacement supercharger. The Roots blower is a common type of supercharger found in many industrial applications, like school buses and in Jackson Racing supercharger kits. A Roots blower has rotating, intermeshing lobes that displace a greater amount of air than the engine's displacement for every revolution of the motor. This is due to both the blowers pumping volume and stepped up rotor drive ratio (which is usually around 2–3 times crankshaft speed). This greater displacement is how the Roots blower achieves a positive manifold pressure, usually about 6–8 psi of manifold pressure. The main advantage that a Roots-type blower has over a turbocharger and a centrifugal supercharger is immediate and proportional response to the throttle as the blower will always pump close to a given constant volume of air over the engine's displacement no matter what the rpm is. There will be slight variations in boost as the blower's volumetric efficiency improves with rpm and the engine's volumetric efficiency varies throughout the powerband, but for the most part, the boost pressure will remain pretty constant. This makes a Roots-blown motor feel like a big displacement version of the same motor. Because of these characteristics, a Roots supercharger is the king of driveabilty. These are perfect characteristics for a daily driver or for those who value low-end grunt. The low-end grunt is welcome in the high revving VTEC B16A and B18C motors.

This cutaway of an Eaton Roots-type blower, as used in Jackson Racing's kits, shows one of the compressor's lobes. The lobes intermesh, spin and positively displace air, forcing it into the manifold where it is compressed and gains positive boost pressure. The Eaton is one of the most efficient Roots blowers made. The brass valve that looks like a throttle plate on the left side of the blower is an internal bypass valve that disables the blower under part throttle operation.

These power delivery characteristics make the Roots blower ideal for the driver who values a very linear response of the motor, such as an autocrosser who needs smooth predictable power from very low speeds to help throttle steer through tight cones, or a driver who wants a tractable, predictable street car with power on demand to quickly end a stoplight encounter.

Eaton—Typically, an Eaton Roots blower, the most popular type, can give a power increase of as much as 40 percent, a very impressive power gain for a single bolt-on piece. The Eaton blower is a well-engineered, proven unit. Since it is used as an OEM part for several different manufacturers such as General Motors, the Eaton has undergone extensive testing for reliability and durability. It is perhaps the most reliable supercharger on the market. The Eaton blower has an adiabatic efficiency of over 60%. Adiabatic efficiency is the difference between how much the discharge temperature of the supercharger is increased and the theoretical pressure gain over compressing air to the same pressure as expressed by the ideal gas law, PV=NrT. This is a big improvement over older Roots superchargers that had efficiencies of around 50%.

Jackson Racing—The well-engineered Jackson Racing Eaton supercharger kits are Roots blowers. Jackson Racing's kits are a big bang for the buck producing lots of power

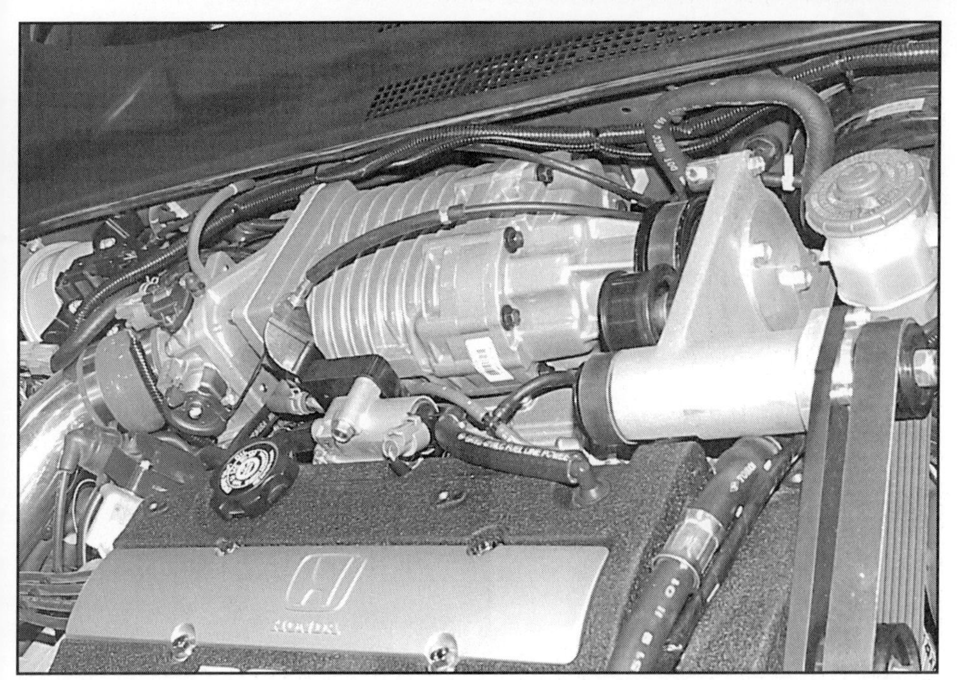

Here is one of Jackson Racing's H22 Prelude supercharger kits installed in a car.

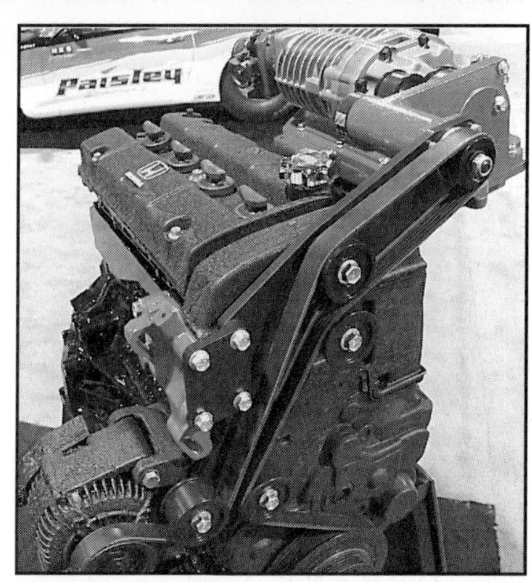

The Jackson Prelude kit out of the car on a display engine. Note the attention to detail and the sano engineering in this kit.

Jackson's B16A kit will work on the Civic Si and the Vtec powered Del Sols.

for the money. They are CARB certified also. A Jackson Eaton–powered B18C5 Type R Integra kit makes up to 225 nice broad predictable hp at the front wheels. The Civic Si kit makes 192 hp and 137 ft-lbs. of torque to the wheels. Greater gains can be expected when items like headers, cat-back exhaust systems and cold air intakes are added to the supercharger system.

Jackson Racing's kits are cleanly engineered and complete with everything you need to install on your car. The kits include a well-integrated OEM-style cast intake manifold that the supercharger mounts to, with all of the brackets, pulleys and duct work needed for a clean installation. Jackson kits are compatible with all power accessories like A/C and power steering, and they keep all of the emissions control components intact.

Fuel enrichment under boosted operation is handled by a boost-proportioned step up regulator that pinches off the return line in the fuel rail to increase fuel pressure in proportion to the boost. An included high-volume fuel pump is standard in most applications where additional fuel volume is needed. Ignition timing is retarded about 6 degrees to avoid detonation. Since the Roots blower has decent boost at low rpm, the only driveabilty issues from the retarded timing is a slight stumble at part throttle off of idle.

At light loads and partial throttle, the Eaton blower features an internal bypass valve, which keeps the supercharger from going into boost when there is a vacuum in the intake manifold. This reduces strain on the engine and helps fuel economy at part throttle by effectively decoupling the supercharger under light load

The internals are exposed in this cutaway of a Powerdyne centrifugal supercharger. Note that centrifugal superchargers are much like the compressor side of a turbocharger.

conditions, virtually eliminating its drag. When the throttle is applied and the manifold vacuum drops, the valve closes, allowing the supercharger to make boost.

Jackson Racing also offers a water injection option. This uses the Aquamist system that works so well for turbocharged cars. See Chapter 9 on Turbocharging for more details on the water injection system. On a supercharged car, the water is a help, allowing much or all of the retard to be taken out of the timing by reducing detonation. The water helps prevent detonation in the cylinders by slightly retarding combustion. It also helps cool the intake charge considerably on the non-intercooled Jackson system. The water is injected right before the intake to the supercharger where it helps reduce the temperature rise caused by the supercharger's compression of the air by using water's high latent heat of evaporation. It is also possible to run more boost with a different pulley with water. *Sport Compact Car* magazine got over 200 hp to the wheels on a Jackson supercharged B16A Civic Si after water injection was added and the engine was tuned to take advantage of its detonation suppression and intercooling capabilities.

Perhaps the best feature of the Jackson Racing kit is its availability. Currently Jackson Racing has kits available for most of the Civic line from 1988 and up, including hybrids with the Japanese Civic ZC motor, the Prelude and the 94-and-later Integra, including the Type-R. Other supercharger companies have limited availability to a select few models of the Honda line.

Some individuals have used the Jackson Racing kit but shunned Jackson's fueling method of using a boost dependant fuel pressure regulator, and retarding the engine's distributor, instead opting to use a programmable stand-alone engine management system. When properly programmed, a stand-alone ECU with a Jackson kit can see additional impressive gains, as much as 15 more hp to the wheels with crisp low-speed driveabilty. Gains in power are made across the board. People have had great success in using the Hon Data and Zdyne user programmable ECUs as well as the Electromotive TEC II, although there is no reason why other stand-alone systems could not be used.

Centrifugal Superchargers

Centrifugal superchargers, like the Vortec and Powerdyne kits, are also capable of producing respectable amounts of power. This is due to their efficient centrifugal compressors. Centrifugal compressors are more thermally efficient than the Roots blower, which is the main reason why the centrifugal supercharger makes more peak power than the Roots. Efficient or not, due to its direct-coupled positive displacement design, the Roots supercharger is still the king of driveabilty and throttle response because of its ability to make low rpm boost.

A centrifugal supercharger, like a turbo, must be spun to a critical speed before much boost can be produced. Centrifugal superchargers are basically a compressor section of a turbocharger, driven through a step-up gearbox (Vortech) or a belt drive (Powerdyne), to the crank. The step-up drive is necessary because centrifugal compressors must spin very fast, much faster than the engine will turn and faster than a Roots-type blower, in the order of 30,000 to 60,000 rpm.

Here is the step-up belt drive of the Powerdyne supercharger. Since centrifugal superchargers must spin very fast to make boost, it is necessary to step them up to spin them much faster than the crank speed. The Powerdyne uses a belt drive, which is supposed to be quieter than the gear type transmission that the Vortech superchargers use.

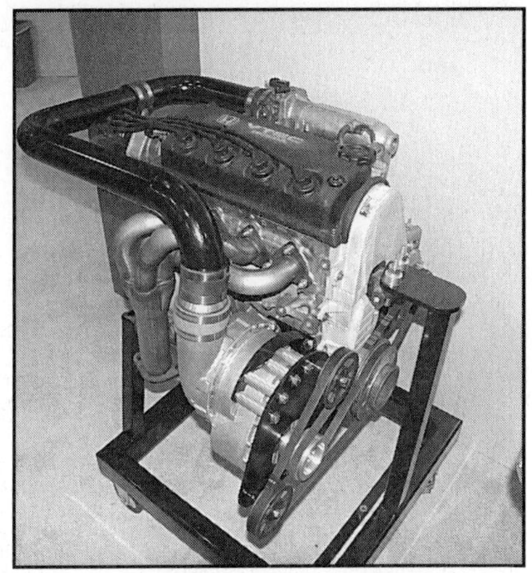

Here is Powerdyne's kit for the workhorse D16 Civic motor. Powerdyne says this kit produces 170-wheel hp on your typical otherwise stock Civic.

Disadvantages—One of the disadvantages of centrifugal superchargers is that the centrifugal compressor by nature works best over a rather small rpm range. To prevent overspeeding the compressor at high rpm, you must select a step-up gear ratio that will not spin the compressor too fast at the engine's maximum rpm. Because of this, the compressor is spinning nowhere near its optimal boost-producing rpm when the engine is at lower rpm, so it does not have much low rpm power boost.

Since the crankshaft directly drives the centrifugal compressor, care must be taken not to overdrive the compressor into surge at high engine rpm. Surge is where the air backs up in the compressor and oscillates violently back and forth in the compressor wheel. This happens when the engine's swallowing capacity is exceeded by the compressor's output, causing the air to back up in the intake tract. The mechanics of surge is when the pressure after the compressor exceeds the energy of the axial velocity component of the compressor wheel's output, which causes the airflow in the compressor wheel to back up. The airflow backs up, the pressure after the compressor drops and the airflow resumes. In severe surge, this can become a violent oscillation that destroys the thrust bearing of the compressor and even causes mechanical failure of the wheel. To prevent surge, the compressor step-up speed in the blower's gearbox must not be too aggressive as to drive the compressor into surge.

Care must also be taken not to overdrive the compressor wheel into choke. Choke is when the inlet vane tip speed of the compressor wheel exceeds mach or the speed of sound, which causes a marked drop in flow. Because of the gearing needed in the supercharger's gearbox to avoid this during the engine's operational rpm, a centrifugal supercharger builds boost slowly in proportion to the engine rpm, usually reaching maximum boost and power at redline. So, the faster you spin the motor, the more power you get. This can make a centrifugal supercharger feel somewhat laggy and less responsive at low speeds than a Roots blower

This gradually increasing boot curve can make a centrifugal supercharger feel somewhat laggy and less responsive than a Roots blower at low rpm, even though they make more peak hp. Some people feel that a centrifugal blower feels like a super VTEC kicking in. In fact, on a B16A Civic Si, the fat part of the Vortec's boost curve hits just as the high rpm cam lobe is engaged by the VTEC system, giving an impressive kick in the back.

Advantages—On a traction limited FWD vehicle like a Honda, this sort of late and smoothly gradual power delivery can sometimes be advantageous, making the car easier to drive at the limit, especially on a road course. It can also aid drag strip

Here is a close-up view of the Vortech Civic Si B16A kit. Note the water to air intercooler and the Jackshaft, which is needed because Honda engines spin counterclockwise, the opposite of most engines and the fact that there is little room toward the driver's side of the engine.

launches, an important factor if you run on street tires. A Vortech equipped Civic Si can easily run low 14 second quarter-mile times on street radial tires.

Another advantage that a centrifugal supercharger has is the ease in adding an intercooler into the system. Since the supercharger is a separate unit from the engine, it is easy to add an intercooler into the pipework plumbing the blower to the engine. With a Roots blower, the supercharger assembly is usually cast into a special intake manifold, making it difficult to package the intercooler. It is hard to plumb an intercooler into the Roots blowers long triangular discharge vent in any case.

Vortech—The Vortech supercharger kit has an optional water-to-air intercooler. As an example of the awesome power a centrifugal blower is capable of when intercooled, an intercooled Vortech kit can make over 277-wheel hp and 175 lbs-ft. of torque on a B16A at only about 8 psi of boost. With headers, a cold air intake and a cat back exhaust, even bigger gains can be produced. The Vortech V-5 supercharger used in the Si kit has a 73% peak efficiency over a broad range. This higher efficiency means that the discharge air from the compressor will be cooler and the compressor will drain less power from the crankshaft.

Vortech has a Civic Si kit available at this time with a GSR Integra kit and standard Civic kits becoming available by the time you read this. CARB approval is also pending and should be done to make these kits 50-state legal by the time you read this. The Vortech kit features a shaft drive, which is necessary to mount the blower where there is more room. Fuel management, like the Jackson kit, is also by a high volume fuel pump and a boost dependant fuel pressure regulator. A huge bypass valve vents excess air to the atmosphere at partial throttle, taking much of the blower's load off of the engine when just driving normally. The Vortech kit is also compatible with all of the engine's power accessories and emissions equipment.

Short of a turbo, the Vortech kit is perhaps the biggest power adder you can bolt on your car in one single shot.

Powerdyne—Powerdyne is currently producing a centrifugal blower kit for the D16-powered Honda Civic. A Powerdyne supercharger is very similar to the Vortech except the step-up gearbox uses a belt drive instead of gears. The advantage of this is quieter operation as the gear drive in Vortech superchargers has a distinctive whine.

Powerdyne is claiming an increase of about 80 hp to the wheels with 8 psi of boost with their Civic kit. This would be around 170 hp to the wheels on a Civic.

Comptech—Comptech is offering a Whipple positive displacement blower kits for the NSX. The Whipple uses two intermeshing screws instead of the lobed impellers that a Roots blower has; this is called a Lysholm supercharger. Some of its advantages are a higher efficiency than the Roots-type blower although it is difficult to find compressor maps for it to quantitatively say how much more efficient it would be. Whipple claims 75% efficiency, which is outstanding. This high efficiency is a great help when these types of superchargers are difficult to integrate an intercooler into. The NSX kit produces 367 hp

Comptech also offers a Paxton

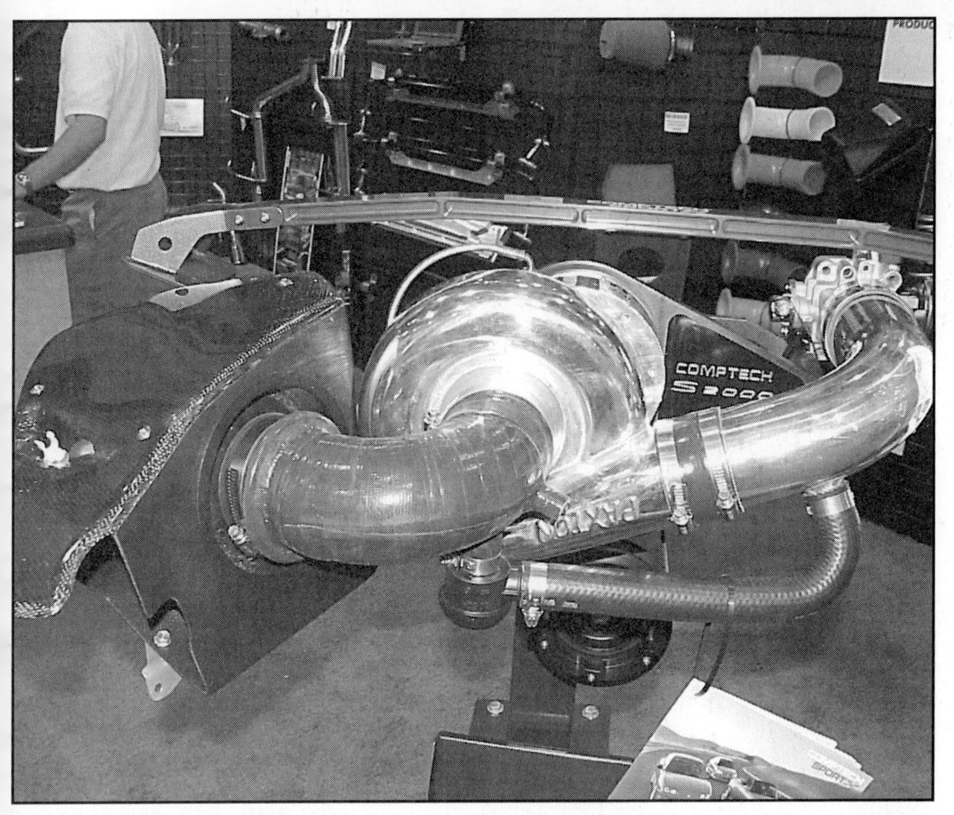

Here is a look at Comptech's impressive Paxton supercharger kit for the S2000. Although most tuners are having trouble extracting significant power gains from the hyper F20C1 S2000, the Comptech kit belts out an amazing 340 hp.

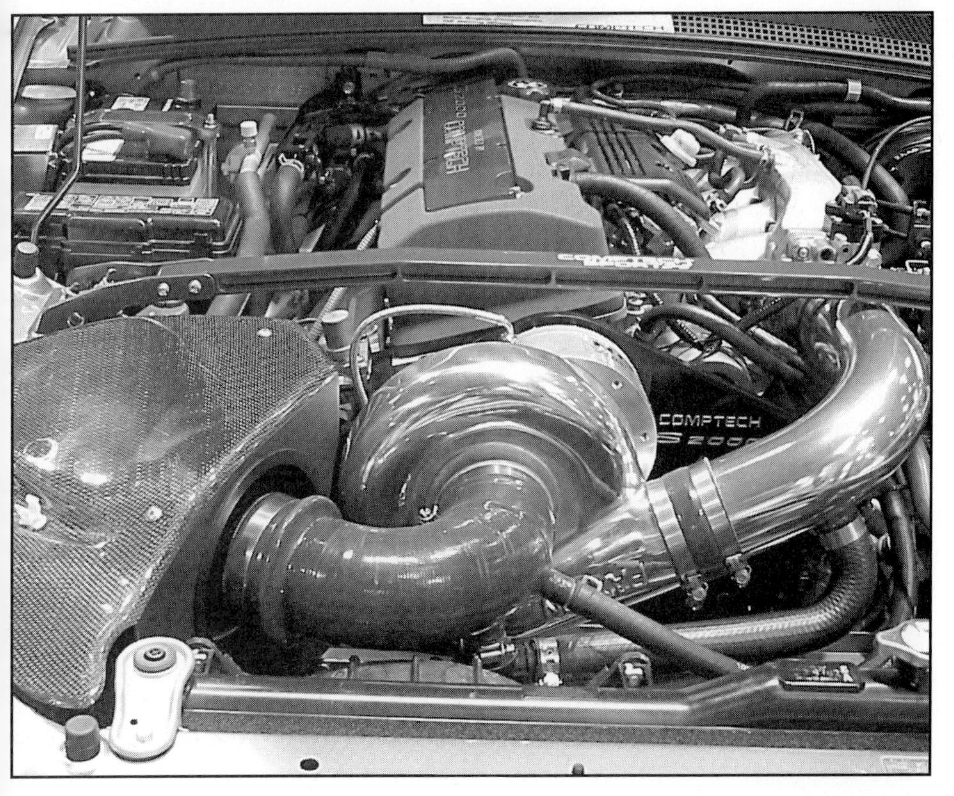

Here is the S2000 kit installed in a car. Check out the trick carbon fiber cold air intake/air box.

centrifugal blower kit for the S2000. The Paxton is a centrifugal supercharger like the Vortech. Paxton does not make any efficiency claims for its supercharger. The S2000 kit makes an impressive 340 hp making it the biggest power added presently available for the S2000.

Both of these kits are CARB approved, a tremendous advantage if you need hassle-free registration in states that have a smog check.

In short, if you are not a racer looking to make the ultimate power and are trying to get a good tractable gain out of your street machine, then a supercharger could be your best bet. The Jackson kits have the widest application range. In a nutshell, Roots positive displacement blowers have the most bottom-end power, proportional throughout the powerband, giving the ultimate in driveabilty. Centrifugal blowers make the most peak power, the power coming on gradually.

9 TURBOCHARGING THE HONDA ENGINE

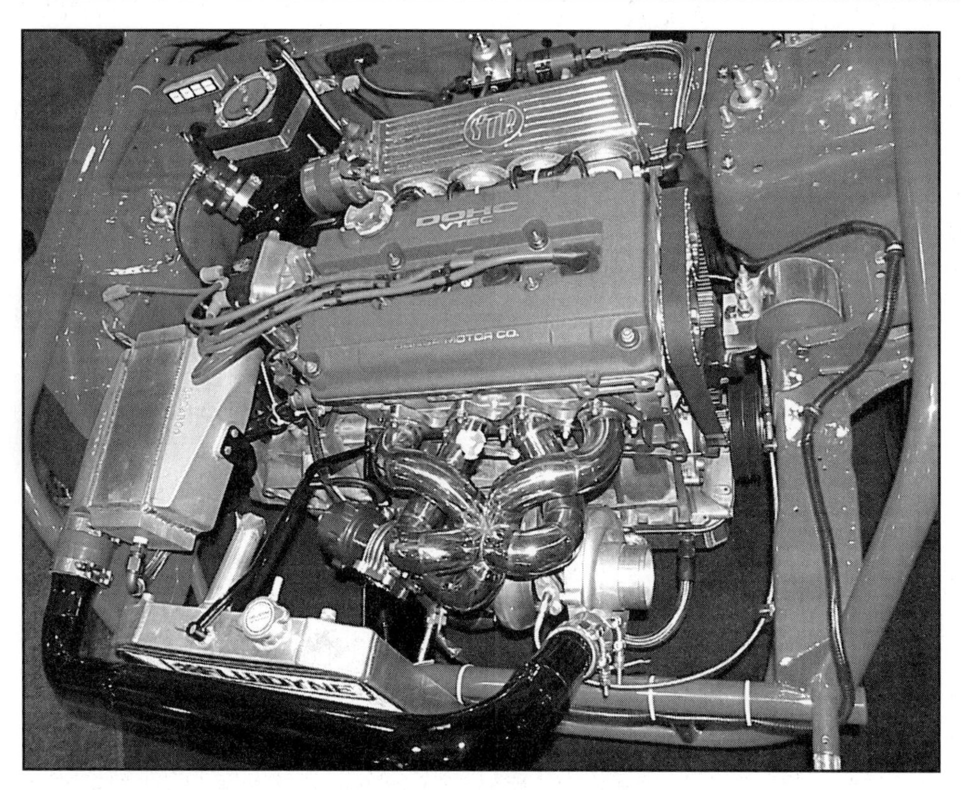

Gary Gardella's Quick Class racer features a Kiwi equal length runner manifold, a Tial wastegate and blowoff valve and a Spearco water-to-air intercooler, all in a very nice package. If you want to play with the big boys, this is what it takes.

The prime device that is largely responsible for the incredible power that stock block Hondas are now generating is the turbocharger. The turbocharger is what enables stock block B18C motors like Lisa Kubo's and Ed Berganholtz's to generate over 650 hp on gasoline from only about 1800cc of displacement. That is only 110 cubic inches for those of you who are metrically impaired. To put this into perspective, if a 350 Chevy could generate the same power per cubic inch, it would have over 2070 hp! I cannot think of too many small-block Chevys that can do that without the aid of exotic oxidizer-laced (read nitromethane) fuel and lots of supercharging. How do turbochargers get so much power? How do you pick a good turbocharger for your application? These are the topics we will address in this chapter.

Since the turbocharger is the one magical device that has single-handedly revolutionized the Honda performance movement and is a complicated enough subject to warrant a book of its own, we have devoted quite a bit of space to it. We might even go as far as saying that the turbo is responsible for the popularity of the compact car performance craze because it put the small motor on a nearly equal footing with the big domestic V-8 motor.

BASIC TURBO THEORY

Basically, a turbocharger is a very simple device. The turbocharger uses a turbine that is driven mainly by the hot expanding exhaust stream to power a centrifugal compressor. The centrifugal compressor pressurizes the intake charge, increasing its density. When more fuel is added to this more oxygen rich, dense compressed mixture, it has more explosive potential energy to be released during the combustion event of the power stroke. By increasing the amount of fuel and air ingested by the motor, a turbo is literally like packing a larger engine into a smaller package. The exhaust stream that drives the exhaust turbine contains a lot of thermal, sonic and kinetic energy. The turbo uses all of this wasted energy to drive the turbine and power the compressor. That is the beauty of a turbocharger. The turbocharger is recovering wasted energy from the exhaust stream to do work, i.e., compress the intake charge to increase its density, energy that is normally just dumped out the exhaust pipe. Since a typical 4-stroke gasoline-fueled, spark ignition engine found in your typical automobile wastes over 30 percent of the energy contained in the fuel right out of the tailpipe, any device that can use this energy to perform a useful function is a godsend.

This use of waste energy is where a turbocharger really shines over other

Here is a partially disassembled turbocharger starting, from left, the exhaust housing, the CHRA (center housing rotating assembly), and the compressor housing.

types of forced induction like supercharging. Because of this use of wasted energy, the turbocharger is capable of producing more power more efficiently than any other form of mechanical power adder yet devised for a 4-stroke engine. Prime examples of the awesome potential of turbocharging are the engines from Formula One's pre-restricted turbo era in the early '90s. These engines produced so much power that they were beyond human control and deemed unsafe. At this point of racing history, the development of the 4-stroke engine was at its peak. A typical 1500cc (this was the FIA displacement limit for turbocharged motors at the time) motor in qualifying trim made over 1500 hp. This power level is a confirmed fact from contacts in the racing industry, from engineers working for a smaller, less successful F-1 team. If a backmarker team's qualifying motor made 1580 hp we must wonder how much power the all-conquering Honda F-1 engines made? Imagine your D15-powered Honda DX making 1500 hp!

TURBO DISADVANTAGES

As awesome as they are, turbos are not the perfect device. Turbos do have some disadvantages to them. The main issues as far as modifying your Honda are turbo lag and backpressure in smaller applications. Most of these issues can be addressed or have been addressed through good sizing of the turbo for the application and through advancing technology in turbo design.

Turbo Lag

Being a turbine-powered compressor driven from the exhaust stream, the turbocharger suffers from what is known as *boost lag*. This is because the turbo is a free-floating device, not mechanically coupled to the engine, and there is a delay from when the throttle is pushed down to when the engine kicks out enough hot exhaust gas to wind the turbine and compressor up for sufficient boost. Typically, this is only an issue below a critical rpm limit where there is not enough exhaust gas volume and energy to spool the turbo up. This can be controlled to a large degree by the design of the turbocharger and how it is installed in the motor. When operating above a certain rpm, the engine produces enough exhaust gas volume for a turbo to spool almost

Here is a close-up of Turbonetics' single ball bearing center section turbo. This system helps speed spool-up.

instantaneously. Since the typical performance turbo spins from around 80,000 rpm when making full boost, to as high as 180,000 rpm for some small frame turbos, it is no wonder that the lag time to spin the turbo up to these astronomically high rpms must be strongly considered when designing a turbo system. Most people think that turbo lag is solely due to the inertia of its rotating parts but lag is caused by a combination of insufficient gas flow at low rpm, inertia and mechanical losses. Insufficient gas flow at low rpm is the biggest contributor to lag, however.

Early turbochargers suffered from a great deal of lag and had pretty poor throttle response, limiting their use to mostly superspeedway racing in USAC Indy cars. With modern turbos, huge advances in compressor and turbine aero design using computer fluid dynamics analysis tools like CFD have much improved efficiency. The advent of advanced CAD packages has been used to optimize wheel mass and reduce its lag-producing inertia. These design improvements have largely overcome the lag problem in the last few years. Other recent innovations such as ball bearing center sections, lightweight compressor and turbine wheels have been used to reduce lag. Low mass compressor and turbine wheels made from exotica like carbon reinforced plastic, ceramics and titanium aluminide are currently in limited production or are being experimented with.

With these engineering advances in turbocharger design and with proper matching of the turbocharger to the engine, boost lag can be largely eliminated or at least managed. When discussing turbo lag, generally as a rule of thumb, given the same motor, small turbochargers spool faster than big ones due to their lower inertia. There are a few other factors to spool and turbo design that we will get into later but this is a good generalization. Larger turbos can pump more air and crate more power than smaller turbos, another good generalization.

As a rule of thumb, given the same motor, small turbochargers spool faster than big ones due to their turbine's ability to extract more energy from the exhaust stream at low exhaust flow levels to drive the compressor and their lower inertia from their smaller rotating mass. There are a few other factors to spool and turbo design that we will get into later, but this is a good generalization. Larger turbos can pump more air and have less backpressure, creating more power than smaller turbos, another good generalization. Big turbos have a harder time collecting energy from a low volume exhaust flow and the bigger parts have more inertia so big turbos have more turbo lag.

A useful analogy is a teapot. If you boil water and put a big pinwheel over the spout, so the pinwheel will be driven by the stream coming out of the teapot, it will spin slowly. If you put a small pinwheel in front of the spout and put a small nozzle on it so the stream shoots out, the small pinwheel will spin up to speed faster.

To the enthusiast, the reduction of turbo lag to acceptable levels has been a boon as the lack of turbo lag has been a major selling point for the supercharger crowd. The supercharger crowd's major claim is that because the supercharger is directly engine driven, they don't have any lag. Well, this is only partially true.

Turbos and Volumetric Efficiency, Backpressure Issues

Another interesting effect of a turbo is that although the turbine recovers waste exhaust gas energy from the expansion of the hot exhaust gas, the kinetic energy of the flowing exhaust gas and the acoustic energy of the exhaust gas, the working turbine also causes an increase in exhaust gas backpressure. This increase in backpressure can hinder the engine's volumetric efficiency. A typical streetable turbo system has more exhaust backpressure than boost pressure and the power gains from such systems are due to the increase in the density of the intake charge, not due to increases in volumetric efficiency (volumetric efficiency is the amount of intake charge inhaled during the intake stroke vs. the actual displacement of the cylinder expressed as a percent.) The larger the percent the better. This is because the smaller turbine housings and turbine wheels used to ensure a quick spool-up time are also by nature more restrictive to exhaust flow. We will explain the mechanics of this in more detail a little later. Racing turbos, the latest generation of medium-sized turbos and turbochargers for engines where throttle response is not too much of an issue meaning fixed industrial engines like generators, long-haul trucks and aircraft, have free-flowing turbines that have less exhaust pressure than intake pressure and do have improved volumetric efficiency. This condition is called *crossover* and is what every turbo system designer strives for. In crossover, VE percentages as high as 110% are not unheard of. Unfortunately, some of the design features such as a large exhaust housing A/R and large, high trim turbines that can create a free-flowing turbo can also contribute to turbo lag, something that is not desirable in a street-driven car that needs a wide dynamic powerband.

Backpressure—Excessive backpressure is hard to manage in a boosted 4-stroke engine. Excess backpressure causes what is known as reversion. Reversion is when hot exhaust gas gets pumped backward into the engine during the overlap period, something that can happen if the turbo has excessive backpressure. Reversion can cause the engine's internals to get excessively hot as cross flow of the cool intake charge during overlap is one of the ways an engine cools itself. Hot internal parts can trigger uncontrolled combustion and engine-destroying detonation. Because of this, it is sometimes not a good idea to crank the boost up on an engine that has a small, high-backpressure, quick-responding turbo. This is the kind of turbo that usually comes on a factory turbo-equipped car.

This is also a good reason not to go crazy with a boost controller on a factory turbo'd car(we'll cover boost controllers later on in this chapter). A little more boost, perhaps 4-5 more psi, might be tolerated, but trying for 20 psi could be flirting with disaster. On small turbo'd cars with a lot of backpressure, camshaft overlap should be kept to a minimum. This means that the stock cam usually will work best. To deal with the problems associated with backpressure and reversion, the engine's tuning must also be compromised with richer mixtures and more retarded timing than what would normally be optimal for the best power. Even on full race turbocharged cars with low backpressure turbos, camshaft overlap should be several degrees less, with more lobe separation angle than on an equivalent, naturally aspirated motor, unless physical measurements indicate that the engine is in crossover in the engine's operational range.

The Jackson Racing Eaton positive displacement Roots-type supercharger, shown here installed on a H22 Prelude, is the king of linear throttle response, low-end power and driveability. Its advantage is not total power production.

Because of the backpressure and VE issues, the correct turbo size to the application is very important when designing a turbo system. A small, quick-spooling turbo can be restrictive, causing a great deal of backpressure, reducing VE at higher rpm. This means that it should be limited to lower boost levels. A big, free-flowing turbo can be laggy and unresponsive, making it unpleasant for street driving, but it will produce awesome power at higher rpm. To combat high backpressure and possible reversion, the compromised tuning needed to prevent destruction with an overboosted small turbo will also reduce power. If a small turbocharger is running a backpressure-to-boost ratio of more than 1.8:1, a supercharger has a good chance of outshining it due to compromised tuning, reduced VE and other reversion-related issues. This is despite the negative parasitic drag issues of the supercharger. Fortunately for us, it is easy to design a reasonably unlaggy, powerful turbo system with a backpressure-to-boost ratio of less than this.

TURBOS VS. SUPERCHARGERS

Some low-inertia, low-friction, small turbochargers are capable of spooling up faster, producing more boost at a lower rpm than even a direct-acting Roots supercharger. The Roots is a common type of supercharger found in many industrial applications, like school buses and in Jackson Racing supercharger kits (see Chapter 8). A Roots blower has rotating, intermeshing lobes that move a greater amount of air than the engine's displacement for every

Centrifugal superchargers like this Powerdyne unit, installed on a D16 Civic motor, usually produce more power than Roots-type superchargers but have a narrower powerband. But even the centrifugal supercharger falls short of the power delivery capability of turbochargers. Turbochargers can have better bottom-end power than centrifugal superchargers if sized correctly.

The Vortech supercharger kit uses a centrifugal supercharger.

revolution of the motor. This greater displacement is how the Roots blower achieves a positive manifold pressure. The main advantage that a Roots-type blower has over a turbocharger and a centrifugal supercharger is immediate and proportional response to the throttle as the blower will always pump close to a given constant volume of air over the engine's displacement, no matter what the rpm is. There will be some variation in boost pressure as the blower's volumetric efficiency improves with rpm and the engine's volumetric efficiency varies throughout the powerband, but mainly, the Roots blower is capable of making more boost at the lowest rpm than any other form of forced induction. This makes a Roots-blown motor feel like a big displacement version of the same motor.

Since the turbocharger is a free-floating device not directly coupled to the crankshaft, it is possible to design a turbo that actually allows a proportionally greater percentage of boost at a lower rpm than even a Roots blower, although the Roots still has a slight driveability advantage because of its very linear throttle response due to its direct coupling to the engine.

The other common supercharger found in the automotive aftermarket is the centrifugal supercharger. Centrifugal superchargers are basically a compressor section of a turbocharger, driven through a step-up gearbox to the crank. The gearbox is necessary because centrifugal compressors spin very fast, much faster than a Roots-type blower, in the order of 30,000 to 60,000 rpm. One of the disadvantages of the centrifugal supercharger is that the centrifugal compressor works best over a rather small rpm range. To prevent overspeeding the compressor into a choke condition (where the air velocity inside the compressor reaches mach speed and stops flowing) at high rpm means selecting a step-up gear ratio that will not spin the compressor too fast at the engine's maximum rpm. Because of this, the compressor is spinning nowhere near its optimal boost-producing rpm when the engine is at a lower rpm. Thus, even though the centrifugal blower is connected directly to the crankshaft, it still suffers from power lag, a non-linear power delivery sometimes worse than that of even a fairly aggressively large-sized turbo.

Better Boost Control—Although a turbocharger is a centrifugal compressor with a narrow operating speed range like a Vortech, since it is not directly coupled to the crankshaft, boost pressure and turbo speed can be regulated independently of crank rpm. This gives the turbo a potentially wider operating range than a centrifugal supercharger. To control compressor speeds and boost pressure produced by a turbocharger, a device called a *wastegate* is used. A wastegate is a valve that allows hot exhaust gasses to bypass the turbine when the set level of boost pressure is reached. The wastegate allows the turbo's compressor to remain close to its most efficient operation rpm over a wide range of engine speeds. With a wastegate, the turbo can be sized to come up to the set boost pressure rapidly, then the wastegate opens, allowing the turbo to stay at a constant efficient speed even as the engine's rpm climbs. The wastegate prevents the turbo from overspeeding, creating too much boost, moving into an inefficient area of the compressor map or even exploding due to too much dynamic stress. The regulation provided by the wastegate is one of the reasons why a turbocharger works so well over a broad range of engine speeds.

A good streetable turbo system like a Drag turbo kit has moderate amounts of lag on a Honda B18C, but will have much of its power going by 4000 rpm, and pump out as much 360 hp. Apexi has a turbo kit for the D-series SOHC Honda motor that can pump out as much as 300 hp. Although the hp for existing supercharger systems on the market seems stalled at the high 200 hp level, there are turbo systems that can be easily fabricated for small 4-cylinders that pump out more than 600 hp. Just check out the many 10-second and faster cars that show up to any given import drag event. Turbos have a big advantage when it comes to power production.

Use of Waste Energy

As stated before, turbos are driven by the wasted 30% of fuel energy dumping out of the exhaust stream without parasitic losses on the crankshaft. This use of waste energy is the main reason why turbochargers have a tremendous power advantage over superchargers.

To put this in perspective, even the low boost 6–8 psi street superchargers like the Jackson Racing and Vortech kits take about 10–20 hp from the crankshaft to spin them up. It takes over 800 hp to turn the big Roots blower on a Top Fuel drag racer!

Calculating Power to Compress Air

Here is the long and boring nerd way to add some validity to the power debate. Let's figure out how much power it takes to compress the air to make serious engine power. We will do this two ways: we will figure the differences in intake air temp for a turbo, a Roots supercharger and a centrifugal supercharger. Then we will show how much power it takes to turn the supercharger. It should all make sense once it's laid out. To perform the following math, you'll have to skip ahead to the section on Compressor Mapping on page 147.

First let's calculate Delta T for the various compressors. Delta T is the change in the intake air temp after it is compressed.

Delta T = Intake Absolute Temperature x (Pressure Ratio to the .238 power –1)/ Compressor Efficiency

Let's assume our engine is going to run 20 psi of boost or a pressure ratio of 2.36.

Pressure ratio= boost pressure + 14.7/14.7

The temperature scale engineers use for absolute temperature is the Rankin Scale. On the Rankin Scale, zero degrees is absolute zero. So assuming our intake air temp is 85 degrees, let's call that 545 degrees Rankin.

Let's say our match car is a hot Acura GSR with a B18C motor. By using the compressor matching equations used on page 148, we figure that the B18C can flow about 45.3 lbs/min going full tilt at 20 psi.

So here we go. Let's figure out Delta T for our good turbo first. Let's assume an efficiency of 78%, as there are many turbos that can do that at the given flow and pressure ratio.

Delta T= 545 x (2.36 to the .238 power –1)/ 0.78
Delta T= 158 degrees

A 4-cylinder sized centrifugal supercharger is probably much less than 70% efficient at this point but let's be kind to it and assume that, plugging and chugging gets us a temp of 177 degrees.

Now there are not any current Roots blowers on the market that can support this sort of boost pressure on a small 4-cylinder car but let's

Innovative Turbo Systems' huge 1700 hp GTB100 turbo dwarfs a Garrett T15, weed-whacker turbocharger.

suppose there is and let's be very kind, assuming it will get 60% compressor efficiency at 20 psi and 45.3 lbs/min. of flow. This is not likely but let's be nice to the poor Roots. Plug and chug and we get a Delta T of 206 degrees.

Now Delta T is the difference in temp after being compressed. What would our intake charge temp be for the Roots assuming an intake temp of 85 degrees?

85+206= an egg frying 291 degrees with no intercooler!

Now let's figure out the power required to make the boost from these three compressors. The equation for compressor power needed is.

Power in BTU per minute= Mass Flow x Cp (a coefficient) x Delta T/ Compressor Efficiency

To convert BTU per minute to horsepower divide by 42.4

Power = 45.3 x .242 x 158/78 = 2220 BTU min/42.4 = 52 hp recovered from the exhaust.

So here is the horsepower that won't be taken from the crankshaft but recovered from the exhaust stream by the turbocharger on our GSR.

If you check out the gas power cycle in a thermodynamics book (the PV diagram for you engineers), you need to correct the power equation a little for a supercharger. Since a supercharger adds pressure to one side of the motor and the turbo adds it to both, we need to do an estimation of the power recovered by the supercharger on the intake stroke. The equation is:

(Boost Pressure x Engine Displacement in cubic inches x RPM)/2)/12 x 60 x 550

So for our B18C:

(20psi x 110 x 8500 rpm/2)/396000 = 24 hp

To figure out how much power the centrifugal supercharger takes from the crankshaft, let's plug and chug again, getting a 65 horsepower, subtract 24 hp and you get a crankshaft power loss of 41 hp.

Repeating for the Roots blower gets us a loss of 65 horsepower stolen from the crank.

So by simply reducing our meager data, if you calculate the potential hp of our turbocharged B18C, we will get about 453 hp. This might be around 412 hp from our centrifugally supercharged version of the same motor and 388 hp from our Roots equipped motor.

Now this example has been vastly oversimplified and does not take into account differences in VE, air density differences for intercooler effectiveness or lack thereof, dynamic matching to compressor maps and engine tuning variables that have to be different between the types of compressors. Of these variables, all of them except for VE differences will be in favor of the turbo. This shows that all other things being equal, the turbocharger does have an advantage when it comes to sheer power output. The compressor efficiencies we used for the superchargers were very conservative. In real life the superchargers would probably be much worse at 20 psi. If we upped the boost higher, to higher pressure ratios where turbochargers really shine, the calculated differences would be even greater.

Staying in the Compressor's Sweet Spot

As we stated before, a lot of how a turbocharger behaves is dependent on how it is matched to the engine. Turbos come in many sizes and can be precisely tailored to the engine's displacement and the owner's desire in power characteristics. Since the turbo is not directly coupled to the engine's crankshaft, there is some latency in the throttle response, described as turbo lag elsewhere in this chapter. As explained, this lag can be largely tuned out but it will always be there to some degree. Since the turbo is free floating from the crankshaft, creative sizing and wastegating enables the turbo to be kept in a more efficient range of operation over a wider band of the engine's operating range.

Compressor Efficiency and Flexibility

Another advantage that most turbochargers have over superchargers is improved compressor efficiency over superchargers. Compressor efficiency is the thermodynamically calculated temperature rise of compressing intake air a given amount via the ideal gas law divided by the actual temperature rise of air compressed to the same pressure by the compressor in question. This is also known as the adiabatic efficiency. When talking about compressor efficiency, it is given as a percentage with the higher the percentage, the better. The higher efficiency means that the intake charge will be heated less while it is being compressed. A centrifugal compressor like the one found on a turbocharger is very efficient. Typically the efficiency is at least 70 percent over a broad range of engine operation for most turbochargers with state of the art examples having efficiencies around 80 percent. Typical good turbo compressors found in the aftermarket, like the Garrett TO4E, have efficiencies ranging in the mid 70s. Newer designs like Garrett's GT series can have efficiencies of up to 80 percent. The more efficient the compressor, the cooler the intake charge air will be. In the case of an old-school, less than 50% efficient Roots blower vs. a state of the art turbo, the difference in compressor discharge air temp can be over 100 degrees F at the same boost pressure!

Good turbocharger compressor efficiency is important for reducing engine backpressure on a turbocharged car as well. A more efficient compressor requires less power to compress the air, thus the turbine has to recover less energy from the exhaust stream. With less of a pressure drop required to recover the energy, the exhaust backpressure is reduced and the volumetric efficiency goes up. A more efficient compressor also has less turbo lag for this very reason.

The latest designs of turbos like the Garrett GT series and the latest offerings from IHI have highly efficient compressor and turbine wheels combined with low friction ball bearing center sections. The IHI compressor also features an abradeable lining to the compressor housing. This Teflon-like coating is abraded away by the compressor wheel to allow the tightest possible housing-to-wheel fit. This significantly improves compressor efficiency. These high tech features reduce turbo lag to a large degree without sacrificing flow.

In contrast, some newer Roots blowers like the Eaton have improved designs and higher efficiencies in the low 60% range, a vast improvement from the old 50% efficient bus supercharger, but still much less than the typical turbocharger compressor. The Roots design is less efficient because it has a lot of internal leakage and does not have internal compression, relying instead on external compression in the manifold. The Eaton's improved efficiency over prior Roots superchargers comes from its having some internal compression due to changes in its port timing and rotor geometry over old-style Roots blowers. Modern CNC machining techniques also allow the Eaton to have tighter tolerances and less internal leakage then the older designs had.

The inability to operate with good efficiency at high-pressure ratios on small engines seems to be a major limiting factor for centrifugal superchargers when it comes to making race winning levels of power with 4-cylinders. Most centrifugal superchargers of the proper size for a 4-cylinder are out in the 65% or less Efficiency Island of the compressor map at a pressure ratio of over 2.4:1 or just a tick under 20 psi. The bigger centrifugal superchargers that can get into the higher-pressure ratios are in the surge zone with smaller 4-cylinder motors like ours. It takes more than 20 psi to win in the Quick and Outlaw classes these days.

Calculating Pressure Ratio—To calculate pressure ratio, it is simply atmospheric pressure x boost pressure divided by atmospheric pressure. Atmospheric pressure is 14.7 psi. At 2.4:1 the centrifugal supercharger is nearly off the map, close to the surge line on a small 4-cylinder motor and

The Vortec B16A supercharger kit features a water-to-air intercooler and shaft drive to deal with Honda's reverse rotation engines. This setup can produce some awesome power, but not as awesome as a turbo!

operating in an inefficient part of its operating range where it is heating the air and eating a lot of crank power. This is just not enough boost pressure to make serious quick sixteen winning power with a small displacement 4 cylinder. Many turbochargers are capable of efficiently (easily over 72%) operating at pressure ratios of 3:1 (30 psi) and higher on small engines without surge.

The final negative supercharger issue is the lack of tuning flexibility. On a turbo, you can drastically alter the boost levels by tweaking the wastegate pressure signal with a boost controller. Are you at 15 psi and dump in some race gas? You can crank her up to 20 psi and enjoy the power in a second. Are you racing, running 23 psi and the track starts to hook up better? No problem, turn her up to 27 psi to take advantage of it. With a supercharger, you have to change drive ratios, which is not very practical between rounds.

Vortech is the most efficient of the centrifugal superchargers on the market. One of the reasons why some aftermarket centrifugal supercharger compressors are not as efficient as turbochargers is that typically a small aftermarket company usually engineers them where as most turbochargers are designed by big OEM automotive supply companies and their product is highly engineered. Where Garrett has a whole building full of engineers and a multimillion-dollar test facility, an aftermarket company cannot muster anywhere near that level of engineering resources when it comes to design, testing and manufacturing. Buying a turbo from Garrett, Mitsubishi or IHI means that you are buying a unit with lots of engineering time and optimization, way more than what a typical aftermarket centrifugal supercharger will ever hope to have.

Ease of Intercooling

Another advantage of turbochargers is that it is usually easier to incorporate an intercooler into the design. For some reason, supercharger kits do not usually come with an intercooler. Some, like the Roots blower, usually incorporate the supercharger into the intake manifold, making the placement of the intercooler impractical. Centrifugal superchargers could have an intercooler installed as easily as a turbocharger, but there is a perception in the aftermarket that one is not needed because the supercharger runs cooler. This is not true. Most of the heat generated by any sort of automotive compressor is by the compression of the air. Heat conducted through the turbo from the exhaust housing is just a small gain if you run the numbers. Testing also backs this up.

Better Matching to the Engine

Another reason why turbochargers are such kings of power is that turbochargers are used heavily in the long-haul diesel and commercial stationary-engine market. This market covers a wide variety of applications from small 4-cylinder Japanese panel trucks to huge Caterpillar earthmovers. As stated before, fuel economy is very critical to commercial operators as is precise engineering of the engine's power characteristics. This diversity in application engine size and output plus the market demands for efficiency on a large commercial scale ensures that much R&D money is spent on turbocharger optimization which means that there are many different sizes of turbochargers optimized for a great deal of different

The Apexi Honda D-series motor turbo kit is the most powerful single upgrade available for this motor. It can produce up to 300-wheel hp.

Here is a Greddy CARB-approved Prelude turbo kit.

applications. This makes it easy to come up with some very nice turbo combination for most automotive performance users. The supercharger aftermarket tends to be of the "one-size fits all" variety with perhaps about three size applications to cover the performance car market.

THE TURBO SYSTEM

We are now ready to discuss the parts of a turbo system and what they do. Up to this point we have discussed how a turbo works, and now we will discuss the individual components that make up a system, and their function. Armed with this information, you should be able to choose or design the best system for your ride.

For popular cars like just about any Honda, there are quite a few choices in pre-made turbo kits on the market. Drag, F-Max, Apexi, Greddy, Rev-Hard and a few others have bolt-on kits to turbo just about all the Honda models. However, it is possible to easily make your own turbo system or put together your own full-race custom system. If you want to do that or are just interested in what sort of parts are needed for a turbo system to work, read on.

The Turbo

The most important part of a turbo system is, of course, the turbocharger unit itself. The turbo consists of a compressor housing, the exhaust turbine housing and the CHRA or Center Housing Rotating Assembly, which includes the compressor and turbine wheels.

Compressor—The compressor works as the rapidly spinning, aluminum compressor wheel flings air from its exducer or major diameter into the cast aluminum compressor housing at a high velocity, where it is slowed down and thus compressed by the housing's geometry. At the outlet of the compressor wheel, the airflow has a very high velocity, and therefore the flow contains a lot of energy. However, it does not have a lot of static pressure. The diffuser slows the flow down and converts its velocity, or kinetic energy, into static pressure, or "boost." A well-designed diffuser

A Greddy B16A turbo kit is a potent CARB-approved power adder.

can increase the air's static pressure to a level 30-40% higher than it is at the wheel discharge. Without a good diffuser, a large amount of the air's kinetic energy will simply be wasted. This has a detrimental effect on compressor efficiency. The air getting continually centrifugally flung out at the wheel's periphery creates suction, causing air to be continuously drawn into the inducer or minor diameter of the compressor wheel.

The compressor wheel is bolted to the turbine wheel/shaft assembly. The turbine wheel and turbo main shaft are a one-piece unit, usually a stainless or chrome-moly type alloy steel shaft with the investment cast turbine wheel either friction or electron beam welded to the shaft. Turbine wheels are made of a high temperature alloy like inconel, GMR or Mar-M. Inconel is more temperature resistant and desirable in more extreme applications over GMR, which is intended more for low temperature diesel applications. Mar-M is more temperature resistant than either but expensive. The turbine housing is made of either ductile cast iron or better yet, Ni Resist, an alloy of iron with a high nickel content which resists heat better than regular iron, the best but almost unobtainium housings are made of HK-30, a type of high temperature stainless and inconel. These exotic materials are needed to withstand the thermal stress that the turbine receives from the exhaust stream.

Hot, expanding, pulsating gasses from the exhaust ports are diverted by the exhaust manifold into the turbo exhaust housing where they are directed to the inducer, or major diameter of the turbine blades by the volute of the exhaust housing.

Exhaust Housing—The exhaust housing is a snail shell–like enclosure much like the compressor housing but with no diffuser. In the housing, the area of the gas passage decreases as you approach the turbine wheel. This gets the velocity of the flow up. As the velocity goes up, the angular momentum of the flow about the axis of rotation increases. At the turbine discharge, there is ideally zero whirl to the flow, so that effectively, all of the flow's angular momentum is gone. The change in angular momentum is equal to the work created by the turbine. So, the more angular momentum you start with (velocity you start with), the more power you can extract from the flow. This is how the turbine's drive force for the compressor wheel is created. The spent exhaust gasses exit through the turbine wheel exducer, or the minor diameter of the turbine.

Center Housing—The turbine shaft, which supports both the compressor wheel and the turbine wheel, passes through the center housing, which is typically cast iron. The center housing contains the oil passages; a washer-type thrust bearing and semi-floating sleeve-type bearings. If the turbo is watercooled, the water jacket is cast into the housing. The thrust bearings work just like those on an engine's crankshaft, absorbing boost and backpressure side loads, preventing the compressor and turbine wheels from walking back and forth and eating into the center section housing. The shaft bearings are like those used for the engine's crankshaft, relying on a thin film of oil to work. The interesting thing is that they are allowed to float in the housing, which means that high-pressure oil is pumped to both the inside and outside of the bearing. There is clearance on both sides of the bearing for the oil and the bearing actually spins in addition to the shaft at about 1/10 to 1/4 of the shaft speed. The outer layer of oil is not to let the whole bearing spin, but to serve as a damping cushion, which allows the rotating parts to be less sensitive to any imbalance and to help damp out possible destructive harmonics of the rotating assembly. This helps place the turbo shaft rpm of harmonic convergence or the critical speed above the maximum point of operation rpm.

The lubricating oil is prevented from escaping by a two-part seal system. First a slinger disc centrifugally flings oil away from the seal area on the compressor side and a labyrinth after the slinger keeps oil from the compressor. An iron piston ring keeps boost pressure from entering the turbo's center section. This is called a dynamic seal system. Some turbos have a carbon seal. A donut-shaped carbon block is pushed against the shaft's thrust collar by a spring. This carbon donut seals out the oil. Carbon seals have more drag and contribute slightly to turbo lag. On the turbine side, one or two iron piston ring-type seals keep combustion gas from leaking in from the turbine. The turbine also uses an oil slinger and a labyrinth seal like the compressor does. A lot of smoke from the tailpipe is a sure sign that a turbo's seals have failed.

Ball Bearing Center Section—The latest Garrett ball-bearing turbos, favored for their significantly reduced friction and quicker spool-up time, use dual ball bearings instead of sleeve bearings in a floating cartridge. The bearing cartridge is semi-floated to help with shaft motion damping. To date, this is the best, strongest ball bearing center section on the market. Due to the limited availability of these Garrett pieces to the non-OEM market, Innovative Turbo Systems of Simi Valley is developing their own version of the dual ball bearing floating cartridge, which will be plentiful and available to the aftermarket. Turbonetics also has their own version of a ball bearing center section that uses a single angular contact race ceramic ball bearing to take the place of the thrust and

Innovative Systems has a Garrett-based mid-frame size dual floating ball bearing center section turbo. As we said before, ball bearings greatly reduce friction and lag. This turbo has dual ball bearings on a damped floating sleeve to attenuate vibrations. This is a neat unit.

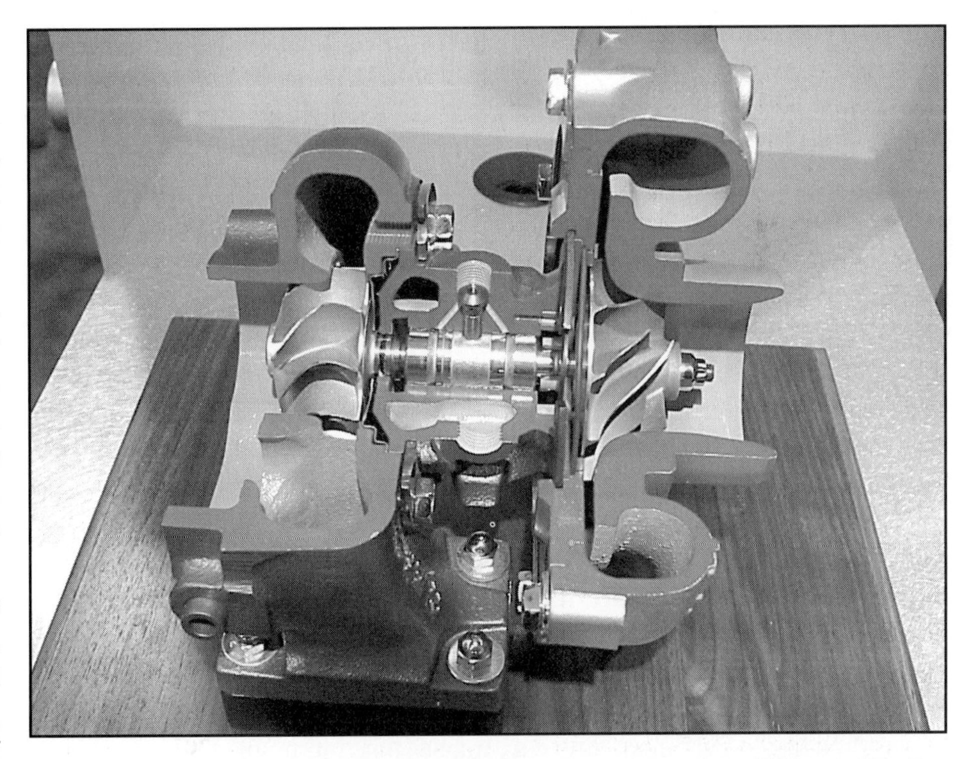

This cutaway of a turbo shows the exhaust turbine, the center-section with its semi-floating bearings and the compressor section when going from left to right.

JoJo Callos's car has a simple 4-1 unequal length runner manifold. Although this is not optimal by design, JoJo is one of the fastest, most successful Quick Class racers in history, so who is to argue! Not all the fast racers have super complicated exhaust manifolds. Note the top mounted Tial external wastegate.

The equal length 321 stainless header by Hi Tech Exhaust on Steff Papadakis's car is a work of art, a fitting piece for the world's fastest FWD Honda.

compressor side floating bearing. On the Turbonetics turbo, the turbine side bearing is still a semi-floating sleeve type. Apexi sells several different ball bearing-equipped IHI turbos. HKS also has several turbos equipped the latest Garrett GT wheels and ball bearing cartridges. We love ball bearing center sections. We have experienced that ball bearing turbos can enjoy 500–800 rpm quicker spool up or can exploit one size larger A/R exhaust housing to gain more flow via better volumetric efficiency with the same amount of lag. These are the major parts of the turbocharger.

Manifold

After the turbo, the next most important item would be the exhaust manifold, which connects the engine to the turbocharger. For most applications, a simple log type or stubby, unequal length runner 4-into-1 manifold works surprisingly well. This is because turbos make a lot of use of the hot expanding exhaust gas to power the turbine and a short runner or log manifold makes good use of this. This is because a log does not dissipate much heat, saving much of it for the turbine's use. A long, runner-tuned manifold is ultimately the best solution for all-out racing, but a log can make a surprisingly large amount of power without the fabrication and thermal cracking problems that a long runner tuned turbo manifold can have. A long runner tuned manifold does not seem to have a huge power advantage until the 400 hp level is passed. Thus said, a long runner tubular manifold is more of a race only setup and a log-type is superior for long-term street use.

Manifold Material—For manifold construction, cast iron is the best for a street setup. Fortunately, cast turbo manifolds are available for all of the Honda's popular motors. If you don't like any of the pre-made offerings, you can make or have your own design custom-made. A good material to fabricate a log manifold from is the cast steel pipes commonly know as Weld-els. Weld-els are used in the industrial world for steam pipes and large hydraulic lines. With their heavy 0.150" wall thickness and their heat-resistant cast steel makeup, Weld-els make an excellent log manifold. When searching for a good fabricated log manifold, look for tig welding and good tight mitering. A Weld-el manifold made like this will be nearly as durable and warp resistant as a cast manifold.

It is good if a fabricated Weld-el or tubular manifold's head flange is cut between the exhaust ports. This gives the manifold a little room to grow and move around as it expands under heat. This helps to prevent cracking and warping. Also look for a slip joint between the wastegate discharge pipe and the down pipe. This is important to prevent cracking in this area that has a lot of thermal expansion movement.

For tubular tuned race manifolds, 321 stainless is the material of choice. This is a strong, heat-resistant stainless alloy. The cheaper 304 tends to crack and the very inexpensive 409 is barely better than regular steel, although it is good for an exhaust pipe. Regular light gauge header mild steel is out of the question here; it just cannot take the heat. Tubular manifolds need to make use of multiple slip joints to make up for the large temperature differential across the long pipes.

On a long runner tubular manifold, the heavy turbo should also be supported by a brace to the engine block. These precautions are important to prevent cracking. A one-piece long runner turbo manifold is asking for trouble and will crack under any sort of extended operation. Any pressure bleeding leaks or small cracks before the turbine in the manifold really kills the turbo performance, so a non-cracking manifold is a must.

If a divided turbine housing is used, the manifold should be configured to take advantage of that. If cylinders 1&4 are fed into one side of the exhaust housing scroll and 2&3 are fed into the other, the distinct separation of exhaust pulses drastically improves turbine efficiency, reducing lag significantly. VE is also improved by reducing cross-cylinder reversion with a divided housing, as it is almost impossible for an exhaust pulse to travel to an adjacent cylinder, as it must go through the turbine first.

The manifold on Miles Bautista's racer is a pulse converter type. By combining runners for cylinders 1 and 4 and 2 with 3 separately into the different sides of a divided exhaust housing, quicker spool up is gained.

Hi Tech Exhaust, the builders of the beautiful manifold on Steff Papadakis's world's quickest Honda also have an equal length street manifold available.

The Wastegate

As mentioned before, the wastegate helps control boost pressure and allows the sizing of the turbo to be so that it can spool quickly but not overspeed. The wastegate works by bleeding exhaust gasses past the turbine wheel when the desired boost level is reached, allowing the turbo to stay at a steady speed, even if the engine rpm and volumetric flow is climbing. Wastegates come in two varieties; internal and external wastegate.

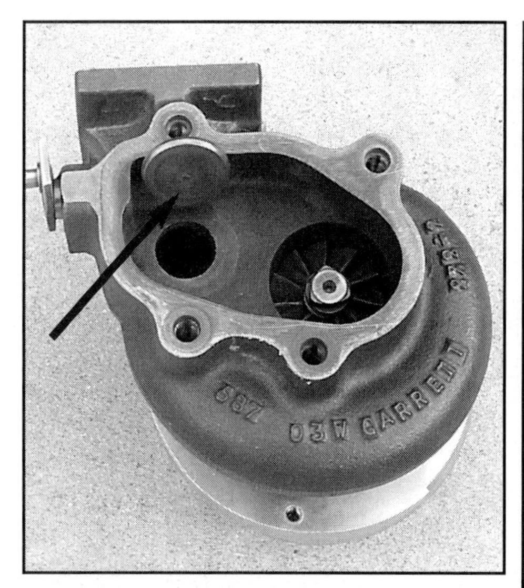

The flapper control valve (arrow) of an internal wastegate is built into the exhaust housing of the turbo. Note how small the control valve is compared to a big external wastegate. This limits the amount of boost pressure and power that an internal wastegate can handle.

Here is the vacuum dashpot that controls the internal wastegate. The dashpot uses boost pressure to open the wastegate by overcoming its internal spring. Note how small the dashpot is compared to the large diaphragm housing of an external wastegate. The small size limits the internal wastegate to reduce boost for lower power applications..

Internal—An internal wastegate uses a passage cast into the exhaust housing that bypasses the turbine metered by a simple flap door connected to a sprung dashpot. The spring in the dashpot holds the wastegate shut until manifold boost pressure fed to the dashpot overcomes the spring pressure and allows the wastegate flap door to open, causing exhaust gas to bypass the turbine. Internal wastegates are cheap and compact, as they are practically built into the exhaust housing of the turbo. For low boost (14 psi and lower) it is hard to beat an internal wastegate for these reasons.

The disadvantage of the internal wastegate is that the diaphragm of the dashpot controlling the wastegate is small and the flap valve is also small. This limits the wastegate flow and also makes the wastegate somewhat slow to respond to manifold pressure fluctuations. Internal wastegates tend to suffer from *boost creep* and *boost spiking* when run at high boost levels because of this. Boost creep is when the wastegate's valve and port are too small and don't flow enough bypass exhaust to control the boost, so the boost continues to climb even when the wastegate is fully open. Boost spiking is when the wastegate is sluggish to respond to a quick rise in manifold pressure and the boost flashes up far beyond the set point. Both boost creep and boost spiking can damage the engine if this momentary, out of control boost condition causes the engine to detonate or if the boost pressure is so great that the engine suffers from mechanical failure.

A slow-acting wastegate with a small control diaphragm and a small valve will also have to open sooner to prevent the boost from spiking, before the set boost is reached. This can slow boost response, adding perceived turbo lag. These disadvantages usually only arise when an internal wastegate is pushed to the flow limit of its flapper valve.

An external wastegate is the much better choice for high-power applications. External wastegates have bigger diaphragms and much bigger control valves, which gives them the ability to respond crisply to changes in manifold pressure, giving the quickest possible response. They can also flow a lot more exhaust, so boost creep and boost spiking do not become a problem. External wastegates also look sexy! Some of the disadvantages of an external wastegate are cost and added exhaust system plumbing complexity as the wastegate discharge should be plumbed back into the exhaust system for better sound.

Contrary to a common myth, if a large wastegate control valve is good enough to control the boost pressure

The Tial external wastegate is a work of art, worthy of just being mounted to a wall. It also happens to be one of the most widely used wastegates on the race circuit.

Steff Papadakis's High Tech Exhaust Manifold shows optimal routing for his Greddy external wastegate.

Steff Papadakis's car also features a separate exhaust system for the wastegate, a good design for maximum power applications.

with no creep, a super large one will not reduce the backpressure on the motor. The wastegate is parallel to the turbine in the manifold and the pressure drop across the wastegate must be equal to the pressure drop across the turbine. The pressure drop across the turbine is solely a function of the turbine's flow and efficiency curves and the power required to drive the compressor.

An uncorked wastegate sounds terrible. Sometimes external wastegates are plumbed separately to their own independent exhaust system. On race applications, this can give more power, as over 40% of the total exhaust flow can be dumped out the wastegate under certain conditions. If the wastegate discharge is dumped back into the exhaust system, it should be at least 18" from the turbine discharge and dumped into a smoothly merged collector. We have seen power gains upwards of 40 hp just by optimizing the wastegate flow.

Blow-Off Valve or Compressor Bypass Valve

The blow-off valve is placed to prevent damage to the turbo's thrust bearings when the throttle is suddenly slammed shut during full boost conditions, such as during an upshift. When the throttle is suddenly closed under full boost, a high-pressure shockwave is reflected back to the compressor wheel, which shoves it violently into the thrust bearing. If this is repeated frequently, severe damage to the turbo will result. This pressure spike can also throw the compressor into surge, which can cause violent internal pressure oscillations and damage the wheels. Severe surge can even fracture the compressor's blades. Spinning at 120,000 rpm, even the slightest unsymmetrical weirdness in a wheel can cause it to explode. At best, the pressure wave will drastically slow down the spinning turbo, causing a loss of power, as the turbo must respool to regain boost pressure after the shift.

To relieve this pressure spike, a turbo kit running more than about 7–8 psi should include a blow-off valve. A blow-off valve is a vacuum-controlled valve with the control diaphragm connected to the intake manifold. When the throttle closes and the intake manifold sees vacuum, the blow-off valve is sucked open and the high-pressure spike is vented harmlessly into the atmosphere. The turbo is spared from the shockwave and maintains much of its spinning

JoJo Callos's car features an HKS sequential blow-off valve.

Here is a close-up of a Tial blow-off valve on Lisa Kubo's car.

The Jun Hyper Yellow Civic features a water-to-air intercooler, a 4-1 equal length manifold with minimal bends and a larger HKS external wastegate.

speed, allowing less drop of boost pressure during the shift.

The blow-off valve is what makes the distinctive whoosh between shifts that many turbo cars have. Many people want to get one just for that noise and many blow-off valves on the market are designed to amplify that noise! Remember, though, that the blow-off valve is an important functional piece of any turbo system, by both helping performance and helping protect the turbo.

Intercooler

When you compress a gas it heats up as expressed by the ideal gas law, which is PV=NrT. If you want to understand the mechanics behind that, then check out a college level chemistry or physics book and read up a bit. This is a physical law of nature and you cannot get away from it.

Depending on the compressor efficiency, the intake charge can be heated to 350 degrees or even hotter as it is compressed. Believe me, your engine does not like ingesting 350-degree air. Reducing intake air temp can significantly increase power. You can experiment with this yourself by plugging different values of manifold air temp into the compressor sizing equations that were given on pages 147–150 and see the direct effect that intake air temp has on power. Hot air has less density and less oxygen, and less oxygen means less power. The hot air also makes the combustion temps higher, promoting overheating and detonation. Hot air is a bad thing for both the motor's life and power.

The intercooler is a heat exchanger whose function is to cool down the intake charge to reduce thermal load and increase intake-charge density. Basically there are two types of intercoolers, air to air and water to air.

Water to Air—A water-to-air intercooler is an intercooler with a heat exchanger in the engine compartment cooled by a jacket of water, circulated by a pump to another remote heat exchanger somewhere in the airflow. Water-to-air intercoolers have a higher efficiency, usually in the 90-percent range, have shorter, less laggy plumbing and are sometimes easier to package, especially for rear and mid-engined cars. For drag and other short duration racing events, a water-to-air intercooler can use an ice tank to cool the intake charge well below ambient temperature. Some of the disadvantages of a water-to-air intercooler are potential heat soaking of the water as the capacity of the water heat exchanger is usually relatively small and additional complexity and weight of the water, extra heat exchanger, pumps and hoses.

Air to Air—An air-to-air intercooler is the most popular type currently in use. An air-to-air intercooler is simply

Steff Papadakis also relies on a big water-to-air intercooler.

Like many leading Quick Class racers, the front end of JoJo Callos's car is dominated by a huge air-to-air intercooler that dwarfs the radiator.

a big heat exchanger mounted in the air stream that cools the intake charge. A well-designed air-to-air intercooler can be up to 85% efficient. The advantage of an air-to-air intercooler is its simplicity and light weight. This makes it a good choice for most use. About the only disadvantage is that the plumbing for it has to be longer with more internal volume, which can contribute to turbo lag. When picking an air-to-air intercooler, it is best to stick with a large frontal area and a core that is only 3" or so thick. 80% of the cooling occurs within the first 2" of an intercooler's core so generally a thin-core intercooler is more efficient than a thick-cored one.

There are also two basic types of air-to-air intercooler cores. Tube and fin and bar and plate. Tube and fin construction is popular with the Japanese intercooler companies like HKS and Greddy. Tube and fin intercoolers are light and are good at exchanging heat but do not have much thermal mass. The mass can be good because it can act like a heat sink at low vehicle air speeds.

Bar and plate construction is favored by American intercooler companies like Spearco and Garrett. Bar and plate intercoolers have a much higher fin density and are even better at exchanging heat. They also have a greater mass so they can act like a heat sink, soaking up heat at low speeds to release it to the air stream once the air speed increases. Although they weigh more, I prefer bar and plate intercoolers for this reason.

As a rule, unless you are minimizing lag by reducing intake tract internal volume, try to run the biggest intercooler you can stuff into the nose of your car within reason. The guidelines of the Spearco catalog are a good place to start in picking your intercooler. The Spearco catalog has very good technical information and engineering data for each of its intercoolers. Remember that total airflow to the engine's radiator should be considered because the intercooler can block a lot of airflow to your radiator. This might be critical for road racing and other long duration events. Sealing the intercooler and giving it directed airflow through ductwork can improve its efficiency up to 40%, this alone can allow you to get away with a much smaller intercooler. To calculate an intercooler's efficiency, go to pages 156–157.

Pipework

All the parts of the turbo system must be connected together with pipework. The important thing to consider here is diameter. Going larger than needed in the intercooler and turbo pipes can have a large impact on turbo lag. Thus it is important to keep the size of the piping as small as possible to do the job.

For setups under 300 hp, the piping diameter should be kept to between 2–2.25" diameter. For up to 400 hp, 2.25" from the compressor discharge to the intercooler with 2.5" from the intercooler to the throttle body works well. For up to 500 hp and more, 2.5" works pretty well. Piping should be constructed with smooth mandrel bends and two-ply silicone hose can

These 304 stainless 3" diameter premade bends from Magnaflow can be used to make fine turbo piping or exhaust piping.

Perforated core mufflers like this 3" diameter Magnaflow are the freest flowing available, making them very desirable for turbo systems.

Lisa Kubo's car has big 850cc/min RC Engineering injectors, a big high-volume fuel rail and a high capacity pump, lines and regulator to feed her 650 hp machine.

make up the couplers. Worm gear hose clamps will work or better yet, aircraft type T-clamps

The exhaust system should be as free flowing as possible. The backpressure of the exhaust system on the engine is multiplied by the turbine's expansion ratio, which is typically around 2. Thus a low 5-psi of exhaust system backpressure quickly becomes 10 psi of power robbing backpressure in the exhaust manifold. That is the reason why turbos respond so well to a free-flowing exhaust. Not only will a low backpressure exhaust give more power but will speed turbo spool up also.

You can almost not go too big with a turbo exhaust. All mufflers should be the straight-through perforated core variety. Small 1500–1600cc engines should have 2.5" diameter exhausts, 1600–2200cc motors, 3" diameter, 2200–2500cc engines should have 3.5" systems. Race systems should be 4". All bends should be smooth, constant diameter mandrel, not crush-bent muffler shop stuff.

Engine Control and Fuel System

Of course your fuel system needs to be upgraded for turbocharged operation. Low boost (10 psi and below) systems often use a high pressure, high volume fuel pump combined with a rising rate fuel pressure regulator. The rising rate fuel pressure regulator increases fuel pressure in proportion to the boost level in a ratio of 1-5 to 1-14. This means 5-14 more psi of fuel pressure for every pound of boost pressure. At higher pressures, more fuel is forced out of the injectors for the same pulse width. Up to 40% more fuel can be squeezed out of the injectors this way. This system is somewhat crude but it can work for lower boost motors. The driveabilty will be less than perfect, but it will work. Interestingly enough, all of the aftermarket supercharger kits use this sort of fuel enrichment.

Japanese "Magic Box"—Another approach is the Japanese tuner magic box. These boxes intercept the signal from the MAF or MAP sensor and allow the engine's ECU to be tuned by turning knobs on a box. The magic boxes can help trim the air/fuel mixture if larger injectors are being run and more fuel is needed under boost and less at idle and part throttle. The Apexi SAFC, and HKS VPC work like this. Magic boxes can work fairly well but certain cars will go into limp home mode if the box is cranked to far, like if you were trying to compensate for really big injectors at idle. Boxes also seem to suffer from some driveability issues and the

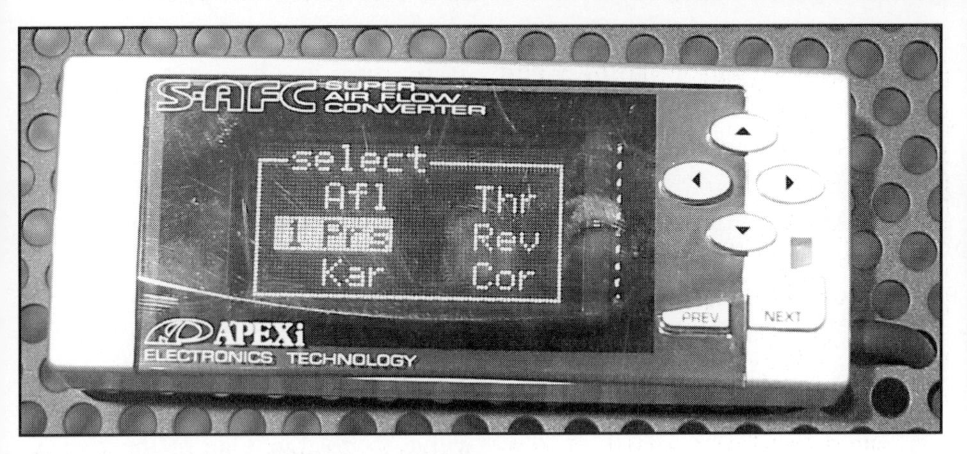

The Apexi SAFC is one of the better magic boxes.

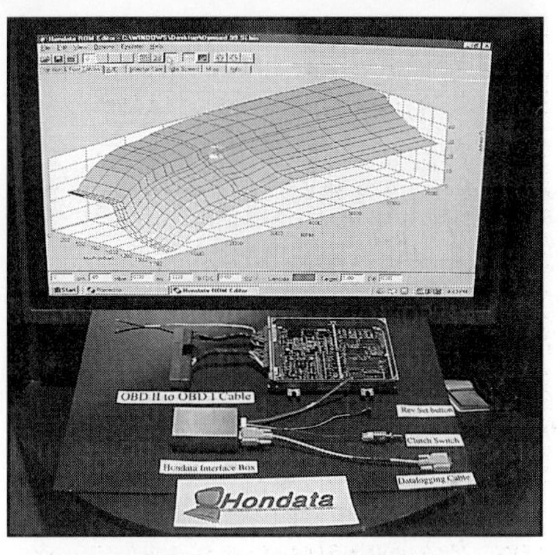

The Han Data ECU is a standard Honda ECU modified to be programmable through a very user-friendly Windows-based interface. This is a very trick system. It is available through G-Force Engineering.

The Accel DFI unit is a popular, low-cost, stand-alone ECU.

tuning is not very precise. Magic boxes also seem to suffer from tuning differences during changes in weather and altitude. There are also boxes like the HKS GCC that can also be used to control ignition advance. Even with these drawbacks, the magic box, if tuned properly will work better than a simple pump and regulator. What can really work well is a magic box used to fine-tune a pump and regulator combo.

User Programmable ECU's—The best way to manage the fuel and spark is a user programmable stock ECU, like the UPRD Hon Data system for Hondas or the Zdyne user programmable stock ECU. These impressive units are modified stock Honda ECU's with a Windows-based user-friendly interface. With these units, you can use all of the stock Honda sensors and wiring harnesses, a big advantage. Stand-alone user programmable ECU like the Accel DFI, the Electromotive Tec-II, Speed-Pro, and the Motec are also good choices. These systems can be programmed with 3D maps for fuel and spark advance, just like the factory ECU, and have most of the driveability and idle control features of stock factory ECUs. They can also be easily tuned to run the big injectors needed to provide the fuel for serious boost. If you are serious about making power and are pushing past 10 psi, these units are the only way to go. AEM has just introduced a powerful stand-alone EFI system that is a direct plug in for the stock Honda wiring harness, using the stock sensors. The unit features data logging, an extremely user-friendly, Windows-based graphical interface and an advanced self-programming feature using a broad band O_2 sensor.

Ignition System

The increased density of the intake charge that the turbo provides increases cylinder pressure and makes the ionization needed to create a spark at the spark plugs much more difficult. This requires a powerful ignition to run without misfiring. Most late-model sport compact cars

The MSD Digital 7 is a very hot ignition system for running very high boost levels.

have pretty hot ignition systems from the factory that can usually support up to 10 psi of boost.

It is critical that the ignition system be in good condition and spark plugs at least one heat range colder than stock be installed. You need colder plugs to help prevent detonation due to too hot of a plug center electrode. New plug wires and distributor cap and rotor are a good idea.

When running a stock ignition, it is usually necessary to close down the plug gap to about 0.30" to prevent misfire under boost. Cutting down the plug gap makes it easier for the ignition to throw a clean spark under pressure. If your gap is closed down to 0.020" and you still have misfire or if you are running more that 10 psi of boost, a high performance ignition system is usually mandatory. In most cases, when turbo'd, a high performance ignition system will help power and driveability, even if just allows you to run the wider stock plug gaps under boost. Cars will run better, especially at low speeds with the wider plug gaps that stock ignitions run under.

When running under pressure, we like the MSD brand of ignition system. MSD seems to have the best balance of ignition power and reliability with good technical and customer service. At minimum, the MSD 6A, Digital 6 or SCI ignition system should be considered; for higher boost levels (above 20 psi) the 7AL or Digital 7 work well. The matching MSD recommended coils and plug wires per application cannot hurt either.

Boost Controls

With no boost control, the spring pressure of your wastegate determines the boost your turbo will reach as the spring determines at what boost pressure the wastegate valve will open. Most wastegates have many different spring weights available for different boost levels. Many wastegates also have an adjustable preload screw for minor boost pressure adjustments by changing preload to the spring.

Most people want to be able to adjust the boost pressure quickly or on the fly. In that case, a boost controller is needed, as changing wastegate springs is a hassle.

Bleed Valve—The simplest boost controller is the bleed valve. A bleed valve simply bleeds a percentage of the reference boost pressure heading toward the wastegate away, fooling the wastegate into thinking that the boost pressure is actually lower. Many people use a cheap aquarium bleed valve to do this. One of these costs just a few bucks.

The trouble with bleed valves is that they can slow wastegate response by buffering the pressure signal. This can be perceived by the driver as additional turbo lag, however this mod is dirt-cheap and it will raise your boost pressure.

Manual Controllers—Better than this are regulator-type manual boost controllers like the HKS VBC. These work by not letting boost pressure reach the wastegate until the pressure can overcome the regulator spring in the boost controller, which is set by turning a knob. You can also make one of these with a small industrial

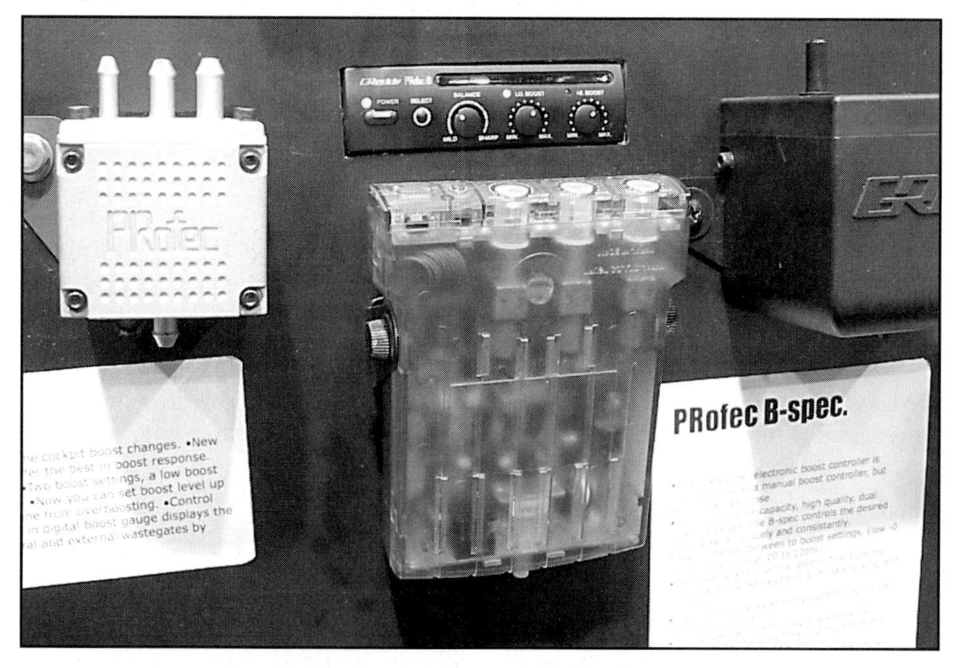

The Greddy Profec B is an excellent non-fuzzy logic electronic boost controller.

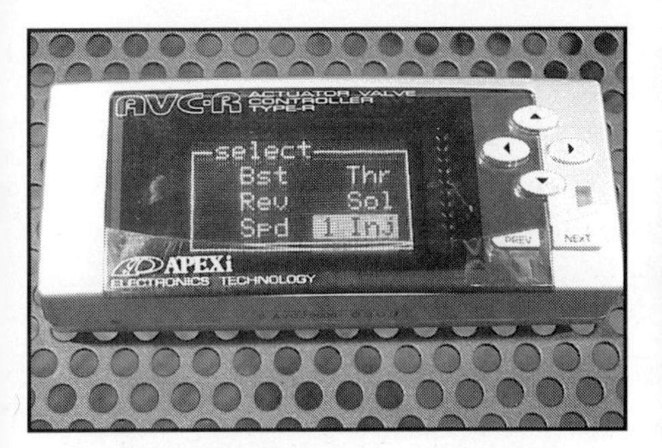

The Apexi AVCR is perhaps the best of the fuzzy logic boost controls. It is pretty complex to operate.

The HKS EVC is a fuzzy logic boost controller while the HKS EZ is a non-fuzzy logic boost controller.

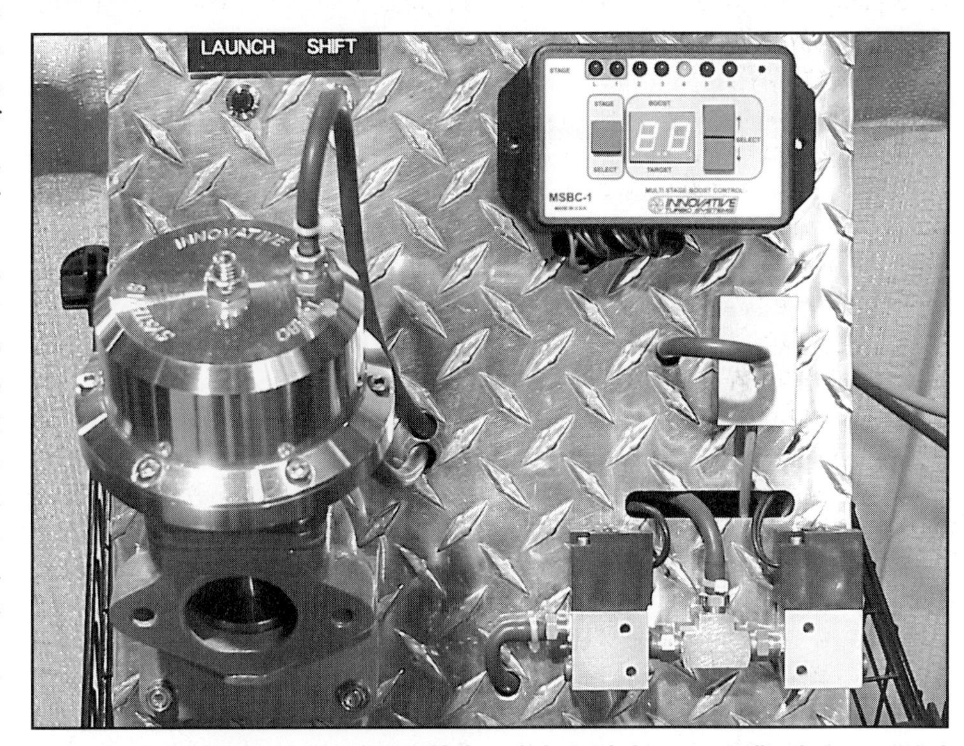

Innovative Turbo Systems has a very sophisticated electronic boost controller that can control boost by rpm, mph and gear position. It should be a boon to a traction-limited FWD turbo Honda.

regulator as sold in industrial supply stores rather cheaply. These controllers allow for crisp wastegate action, as the regulator will hold back the pressure until the preset pressure is reached. Then it will let the pressurized air past, to blow the wastegate open. These types of controllers feel much more crisp and responsive than the cheapo bleeder valve.

A disadvantage is that these controllers should be mounted fairly close to the wastegate for them to work well. Long lengths of hose to the regulator can cause boost spiking and other problems. This means it is usually best to mount the regulator in the engine compartment. That means you must stop and open your hood whenever you want to change the boost pressure.

Electronic Controllers—Finally there is the electronic boost controller. These are units like the HKS EVC, the Greddy Profec and the Apexi AVCR. Electronic boost controllers use either pulse width modulated solenoids or stepper motors to regulate the boost pressure to the wastegate. These controllers also offer crisp wastegate action. The great thing about these units is that the control unit can be easily mounted inside the car at your reach. It is super convenient to adjust the boost pressure with them. I guess that's why a nickname for them is "dial a death." These units make it very tempting to turn her up until she blows!

When buying an electronic boost controller, it is my advice to stay away from the highly touted fuzzy logic units if you have a packaging situation where the boot control solenoid or stepper motor must be mounted far away from the wastegate. For some reason, fuzzy logic controllers do not seem to like these applications. Inconsistent boost pressure and severe boost spikes seem to occur with these units in those situations.

Fuzzy logic boost controllers must go through a learning procedure of driving at wide-open throttle in third

An Aquamist water injection system can be a useful tuning aid for a turbocharged motor. This is an Aquamist high-pressure water pump.

This cutaway JUN B18C turbo motor shows reinforced cylinder sleeves, low compression, dished forged pistons, a metal head gasket, stainless valves and nice head porting.

or fourth gear for a relatively long period of time. This is difficult to do in a crowded urban center like where I live and difficult even on some racetracks. Most fuzzy logic boost controller must be run in this learn mode first or they will not work correctly. This can be scary to do on the street or the freeway. Try weaving in and out of traffic on the freeway with your foot buried to the floor at 100 mph waiting for your boost controller to beep that it has learned your engine's boost curve. Not fun.

Fuzzy logic boost controllers often go wacky and must relearn the engine's response characteristics whenever a tuning change is made or even when a big change in air density is caused by weather or altitude variation. If you like to continuously tinker with your motor, you must be aware of this. Of course this is just my opinion, but more than a few bad experiences formed my view of fuzzy logic controllers, many people swear by them. I have never had any sort of trouble with a non-fuzzy unit like a Greddy Profec B or an HKS EZ.

Cooling System

More power from the motor means more heat. Although the stock radiator is probably OK for short bursts of speed like in typical street use for a hard blast through the gears, autocross or for drag racing, extended high speed running like road racing and time trailing is probably going to give the stock cooling system fits. Adding a surfactant to the radiator coolant like Redline Water wetter or Neo's radiator additive can help as well as running a higher-pressure radiator cap. Ultimately a bigger radiator might be needed. Fluidyne makes a great selection of lightweight, high-capacity aluminum radiators for Hondas.

Contrary to some supercharger companies' marketing people, turbochargers do not inherently make an engine run any hotter than a supercharger will at the same power level, nor is the pressurized air coming out of the turbo's compressor hotter than a supercharger; in fact, in most cases it's cooler.

Water Injection

We have had excellent luck with ERL Aquamist's water injection system. The Aquamist system uses special atomizer water nozzles with a pulsating 120-psi water pump. The Aquamist system can be tuned to produce more power by allowing a more timing advance, a leaner air/fuel mixture and/or more boost pressure. The Aquamist system can also be tuned to add more reliability by preventing detonation and engine overheating at a slight loss of power by changing nozzle sizes and/or a sophisticated 3D computer controller that can be programmed like a stand alone ECU. This makes the Aquamist system a useful tool when trying to turbo high-compression motors like the Honda/Acura B16A/B18C.

Water has a very high latent heat of vaporization, which means that it absorbs a lot of heat while it evaporates. When the fine mist of water is introduced into the intake stream, the intake charge temp is reduced by 20–30 degrees. The water also slightly retards the combustion in the cylinders, damping detonation considerably. Water injection is almost like having a tank of high-octane racing gas all the time.

When running water, if the engine is tuned to take full advantage of the water injection, we have seen up to 40 more hp on a turbo engine on 92-octane pump fuel. Conversely the tuning can be done more conservatively to just get a reduced likelihood of detonation and more reliably and less likelihood for overheating under prolonged high boost. An example where this can really help is extended full throttle top speed run or road course events that last a lot longer than drag racing.

Beware of other water injection systems that have windshield washer-type pumps and non-atomizer nozzles. These pumps hardly have enough pressure to overcome the boost pressure, and a non-atomizer nozzle can have distribution problems.

The Engine

Since we covered the details of the engine's bottom-end in Chapter 4, we won't go into a whole lot of detail here. If 10 lbs. or less boost is going to be run, then most stock bottom ends are OK. Some motors that are a bit less strong like Honda's D-series, should probably stick to 7–8 lbs of boost. Honda's B-series of motors should not exceed 14 psi or so as the cylinder walls and piston ring land strength becomes marginal.

If you are building the motor to take more boost, then low-compression, forged pistons are the logical choice. The aftermarket is full of good connecting rods and there are many acceptable solutions on the market to beef up Honda blocks with block guards, buttressed cylinder liners and weld-in deck plates.

Cylinder head porting helps but does not give proportionally as large a power gain as it will on a normally aspirated motor. Thermo barrier coatings like the excellent product sold by Swain Technologies help reliability and reduce wear. Wire o-ringing and metal head gaskets can prevent boost-induced head gasket blow-out, although this is unlikely until really high boost pressures are hit, in the order of over 20 psi.

Because a turbocharged engine does not require astronomical revs to make power, if detonation is totally contained, a high-output turbo motor can be more reliable than a hyperrevving all-out normally aspirated motor making 100% less power. Revs cause tensile stress on the motor, which is more destructive and harder to contain than the compressive stresses that boost pressure can add.

Now you have about all the information you need to build yourown turbo system, from mild street to full race use. If you are not inclined to build your own system, this information can help you select the best pre-built turbo system for your needs. This knowledge will also give you the ability to intelligently discuss construction details with your turbo supplier and/or fabricator if you are building a custom or full race turbo system. Only turbochargers can give you this sort of flexibility when building your ride. Now go forth and conquer, with a turbo, you are packing the great equalizer of the automotive world! With turbo boost you walk tall, no longer will you fear the dreaded V-8 domestic.

TURBO MATH

Now that you are familiar with all of the basic components of a turbo system, let's delve into a little theory and math that will help you make the ultimate choice. Probably the most critical step in selecting which turbo you should use is figuring out which compressor will work the best. Compressor matching to the engine is also where the most beginner mistakes are made. By using the formula we will soon discuss and looking at various compressor maps, you will be able to closely estimate which compressor will work the best for your intended use.

Compressor Mapping

The math involved in choosing the correct turbocharger is somewhat involved and requires that some technical information from the manufacturer. The most important bit of information is the compressor map. The compressor map is a graph of the compressor's efficiency when the boost expressed, as pressure ratio is the Y-axis of the map and the flow as expressed as pounds of air per minute is the X-axis. The compressor map is two-dimensional and the area efficiencies of the compressor are called islands. The area of compressor surge (area of operation where the air column inside the compressor hits mach speed and the flow starts to oscillate and becomes unstable, basically stopping) is a line bordering the islands on the far left side of the map. Compressor maps are available from the catalogs of Turbonetics, Innovative Turbo Systems and sometimes from calling the manufacturer of the turbos.

To determine if the compressor is a good match for the motor, you must plot the engine's flow requirements over its operational rpm range against the compressor map. Ideally the plot will fall over the map's best efficiency island and stay out of the surge range. The calculations are somewhat involved but if you understand high school algebra, they are not too bad. The more computer literate of you may find this a useful tool and can put these equations into an Excel spreadsheet to make it user-friendly. Here is how it goes, so fire up your calculator and sharpen your pencil.

Step 1—Figure out the maximum level of boost that you plan to run. Most stock Honda motors with proper fueling, on 92-octane pump gas, can handle at least 7–10 psi if enough fuel is provided and detonation is controlled. If you plan on running race gas, figure at least a few more safe psi. If you are building your motor to lower the compression to a turbo friendly 8-8.5:1 compression, figure that you can run 14–18 psi on pump gas and over 20 psi on race fuel. This is all very general and much of it depends on how good your tuning is, how your budget is and how much intestinal fortitude you have. Let's do a compressor match on a B18C.

Assume:

Boost: 20psi is typically what a built Honda motor with race fuel can safely withstand.

Pressure drop across intercooler: Assume 1.5 psi in most cases. If you have a Spearco intercooler, you can look up the pressure drop in their excellent catalog.

From this you can calculate absolute pressure out of the compressor or Pco:

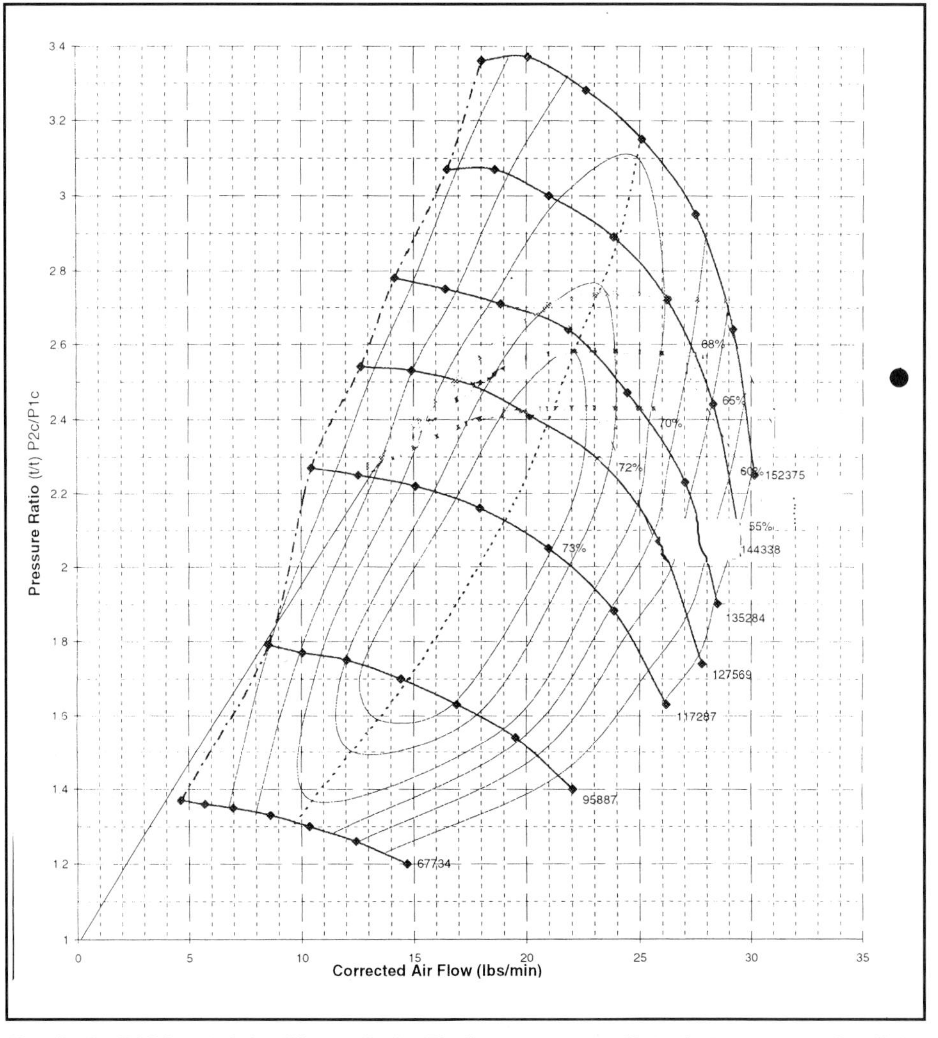

Here is the B18C match for this small trim T3 size compressor. Note that we are totally off the map. Although this compressor will work and is a decent match at lower boost pressures, it is going to blow hot and bothered in this engine at higher boost pressures.

Pco= Boost pressure + Atmospheric Pressure + Pressure drop across the intercooler (assume Atmospheric pressure is 1 bar or 14.7 psi)

Pco= 20+ 14.7+ 1.5
Pco= 36.2 psi

From this you can calculate the pressure ratio or Pr:

Pr= Pco/Atmospheric pressure
Pr= 36.2/14.7
Pr= 2.463

Next guess what the post intercooler temperature might be. We use 130° F as a good starting point as we have found that this is fairly representative of what we have measured with many turbocharged cars with a fairly good aftermarket intercooler.

We will now calculate the approximate density of the air post compressor and intercooler or Di:

Di= Boost pressure+ Atmospheric pressure/ R x 12 x (460+ post intercooler temp)

R= 53.3 (this is a constant from the ideal gas law), 12 is to preserve the inch units in the equation and 460 is to convert degrees F to degrees Rankin (absolute temperature). You don't need to understand this, just always plug these into the equation.

Di= 20+14.7/53.3 x 12 x (460+130)

Di= 9.19 x 10 to the –5th power, lbs/cubic inch

From this we can calculate Mf or the mass flow rate of the engine at the rpm where we want to do the match at:

Mf= Di x Engine displacement in cubic inches x RPM of match point/2 x Volumetric Efficiency

For Volumetric Efficiency we can assume 90%, which is typical for most modern 4-valve DOHC sport compact motors. For an old-school 2-valve motor, you might want to plug in 80%.

Mf= 9.19 x 10 –5th power x 110 (for a 1.8 liter B18C motor) x 8500 rpm/2 x .90
Mf= 47.7 lbs per minute

Now we need to find the corrected mass flow or CMf to get us in the ballpark:

CMf= (Mf x the square root of the sum of, actual compressor inlet temperature in degrees Rankin / 545)/ (Atmospheric Pressure/Corrected Compressor Inlet Pressure)

CMf= (47.7 x the square root of 545/545)/ (14.7/13.95)
CMf= 45.3 lb/minute

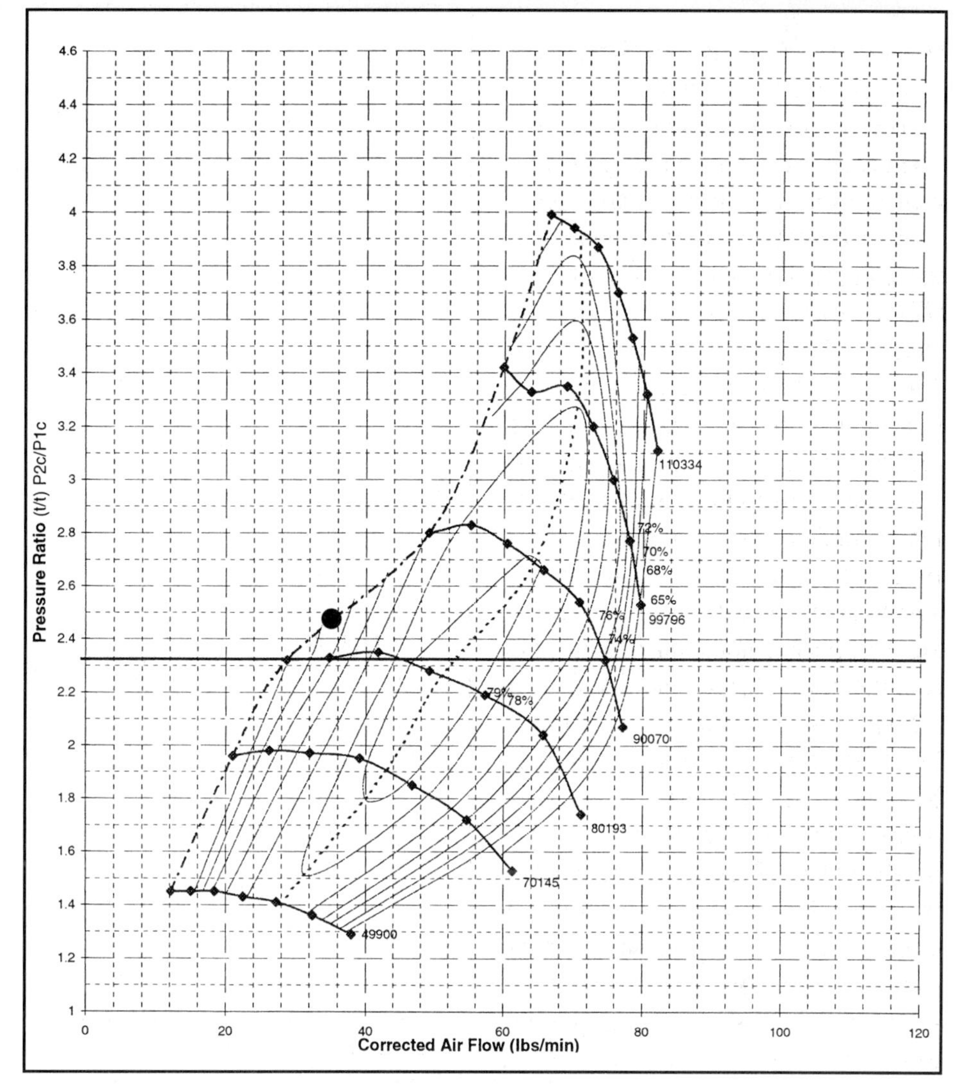

Here is the B18C match for a huge diesel truck compressor to show that you can go too big. Note that the matching point borders the surge line even at the engine's redline! With this particular compressor, the engine operates in the surge zone during most of the engine's operation. This will surely damage the turbo. The engine cannot swallow what this compressor can flow unless the VE gets impossibly way over 100%.

Note: 545 Rankin is 85 degrees Fahrenheit (Rankin is absolute temperature, used in many engineering calculations, 0 degrees R is absolute zero), which also happens to be the standard temp Garrett uses in their compressor maps. So for matching to Garrett compressors, we use this figure. We will also use it for our ambient compressor inlet temp just to make things easier for us. You can subtract whatever the actual temperature you want to use from 545, but we figured that 85 degrees is a good rough estimate of average underhood intake temp and it makes our math easier. Convert your ambient temp degrees from Fahrenheit to Rankin by adding 460 to it. As for the ambient air pressure, it is 14.7 psi. Garrett uses 13.95 psi as the corrected compressor inlet pressure considering the pressure drop across your typical air filter.

So you have your pressure ratio of 2.46 and your mass air flow of 35.9 lbs/minute. Next plot these points on the compressor map that you are considering to see how well your engine matches the compressor map.

If you want to plot more points you can do so by plugging in different rpm values, remembering that no turbo is going to produce 20 psi at idle. I would figure for something between 5000 to 8500 rpm for the B18C. Like we said before, ideally you want to stay away from the surge line and fall across the areas of maximum efficiency. See these three matches below for what a too small, too large and a good match will look like.

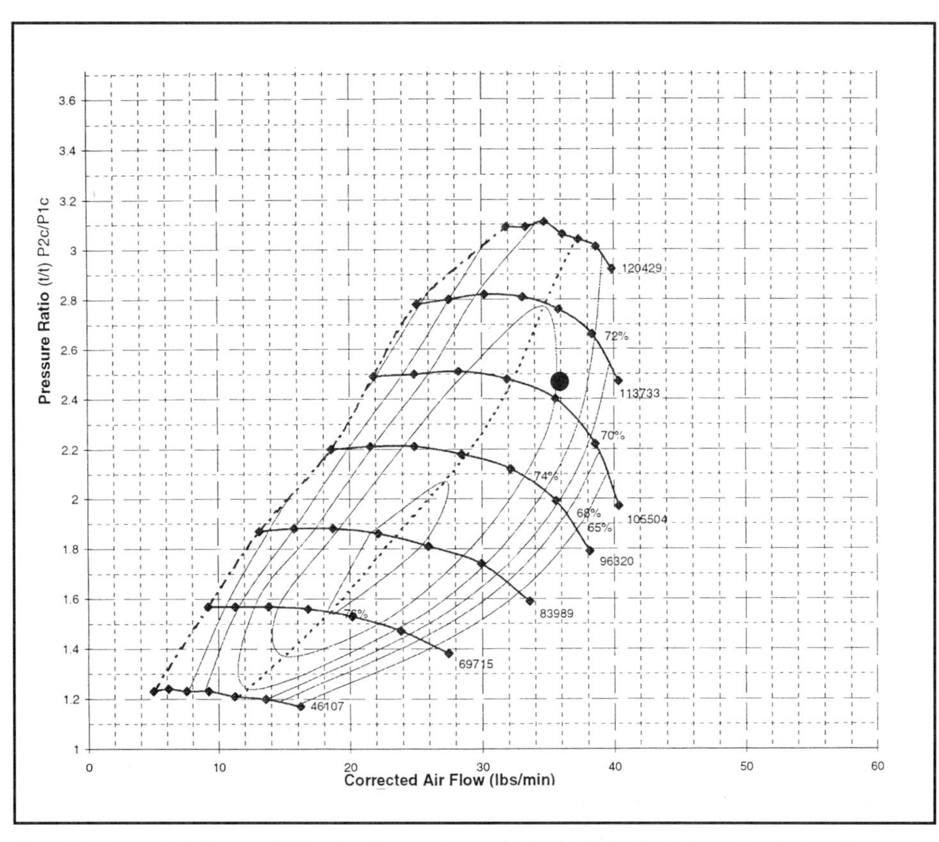

Here is the match for a mid-size T04 compressor wheel. Note that there is plenty of room to the surge line and since the match point is at redline, the engine will operate right across a pretty efficient operational area of the map. Note that at lower boost levels and slightly higher boost levels, the compressor will still be operating in reasonable areas of the map. This is a very flexible compressor selection for this engine and I would say, a good selection.

Avoiding Surge

Perhaps the most important thing to be aware of when matching compressors is avoiding surge. Surge is where the air backs up in the compressor and oscillates violently back and forth in the compressor wheel. This happens when the engine's swallowing capacity is exceeded by the compressor's output, causing the air to back up in the intake tract. The mechanics of surge is when the pressure after the compressor exceeds the energy of the radial velocity component of the compressor wheel's output, which causes the airflow in the compressor wheel to back up. The airflow backs up, the pressure after the compressor drops and the airflow resumes. In severe surge this can become a violent oscillation that destroys the thrust bearing of the turbo and even cause mechanical failure of the wheel. Surge can be felt as a slight fall off in power to a violent jerking when driving. It also makes a thrumming noise that can be distantly heard over the engine.

To avoid surge, turbos that are optimized for higher boost pressures have a narrower B width on the compressor wheel and a shorter, narrower diffuser in the compressor housing. B width is the tip height of the compressor wheel on the exducer or major diameter of the compressor wheel. The narrower B width on the compressor wheel increases velocity of the air leaving the wheel and the shorter, narrower diffuser has less drag losses after the wheel while keeping the important velocity up. These design differences give the air enough velocity and energy to avoid the backflow that starts surge. This is somewhat of a compromise as these design features usually reduce peak efficiency by a few points. Higher pressure ratio compressor wheels also tend to have less backsweep on the blades and thicker blades, both of which also reduce efficiency.

It is also possible for the surge to occur in the diffuser of the compressor housing due to the dynamics of the system. In this case, what is called a ported shroud compressor housing is used. You might have seen these on really big turbos. A ported shroud looks like a really big inlet bib with the inducer bore of the housing inside it. Looking closer, you can see an annular groove around the inducer of the compressor in the housing. What this does is allow air to recirculate around the compressor wheel at higher boost pressures, "fooling" the compressor into thinking it is actually flowing more and helping prevent surge. This also costs you some peak efficiency but preventing surge is much more important. Ported shroud housings are useful in preventing surge at higher pressure ratios.

You want to select a compressor with maximum efficiency for two reasons. The more efficient the compressor, the cooler the intake charge and even more importantly, a more efficient compressor requires less turbine power to create the same amount of boost. This allows the turbo to have less backpressure since the turbine does not have to strain against the exhaust flow to recover the energy to spin the compressor. Less backpressure, the better the volumetric efficiency. This also reduces turbo lag. A good intercooler makes having a super efficient compressor less critical but lower intake charge temp is always good.

With these numbers, you can plot your match point on your typical compressor map to see where in the efficiency range your motor falls in respect to the compressor you are considering. Typically, this amount of mathematical gyrations will get you to easily within 10% of your actual flow. You can get closer with a few more rounds of math but for a street car, this is probably close enough for you, and close enough for you to figure out if the turbo salesman is BS'ing you or not.

How to Calculate Potential HP Output

Once you have figured out your engine's airflow at the boost you plan on running you can estimate your hp output pretty easily. To estimate your potential power output, you can do so easily with this formula:

HP= Airflow x 60/Air/Fuel Ratio x BSFC

Airflow is in lbs/min like we just calculated, 60 is to convert minutes units to hours, air/fuel ratio is self-explanatory, BSFC is Brake Specific Fuel Consumption as measured in lbs. of fuel per hp per hour.

For guidelines, a highly boosted turbo engine running on street 92-octane unleaded pump gas might run an AF ratio of 10.5:1 to hold down detonation. Conversely, a highly massaged drag motor running a specialized high specific gravity turbo fuel might run an AF ratio as lean as 13:1. For estimations on BSFC, a rich tuned pump gas motor might run a BSFC of 0.60. A massaged, tuned to the edge drag race motor on specialized gas might run at 0.45.

So let's pick a crisp state of tune (conservative on race gas) for our hypothetical B18C.

Let's pick a conservative, on race gas air/fuel ratio of 12:1 and a reasonable BSFC of 0.50, this is a very safe state of tune where you won't be close to burning down any motors.

HP= 45.3 x 60/12 x 0.50
HP= 453

So our hypothetical B18C will put out around 453 hp @ 20 psi of boost. If you take the time to play around with the equations, the things the tuner has the most control over are the VE, through different size turbines, turbine housings, exhaust systems, headwork cam combos, etc. and the tuning factors, which include AF ratio and BSFC. The intake manifold temperature can be fiddled with different intercoolers or better yet, go and measure your car's post intercooler intake temp. You can play around with these numbers mathematically and see how they can affect your power and what you can do to extract more power.

How to Size Turbines

Once you have figured out what compressor will work well with your motor and intended use, the next step is figuring out which turbine will match that compressor. Sizing turbines is pretty difficult as I have yet to see any engineering data for turbines available to the general public in aftermarket catalogs or in turbocharger books. The formulas involved are published public domain and involve calculating the power balance between the compressor and turbine to determine the turbine expansion ratio, but I have yet to ever see a turbine map to be able to use this data. For some strange reason, it seems as if the turbo manufacturers want to keep their turbine maps a closely held secret. The math involved in turbine matching is also more involved.

Fortunately, turbochargers are not as sensitive to turbine sizing, as they are compressor sizing. Because of this I have a few rules of thumb to consider when trying to figure out what sort of turbine to use.

1. Try to keep the compressor and turbine wheels to within 15% of their major diameter to each other. One of the semi fads in the aftermarket is to do what is called a hybrid turbo. A successful example of this is a medium-size Garrett T3 turbine with a larger family TO4 compressor. Some of the T3 family have good flowing turbines and they have been teamed up with higher flowing TO4 compressor with good success. This turbo combo has powered more than a few

powerful Honda turbo kits and many 10-second Quick Class racers. This combo takes advantage of the lower inertial mass of the smaller T3 wheel. Most of these combos work pretty well. The compressor in this case is 13% bigger than the turbine, which is still in line with the 15% rule. In fact the T3/T4 hybrid is one of the best-working turbo combinations for a sport compact 4-cylinder. That being said, stay away from the extremes of size mismatches between compressor and turbine wheels.

The reason why you want to keep the compressor and turbine wheels close in diameter most of the time is to keep the wheel tip speeds nearly the same. This avoids efficiency killing slip losses due to the blade tip speed differential. One famous tuner with a very good reputation sells what is basically a large Garrett TO4E compressor wheel hybridized with one of Garretts latest designs, a small GT25 turbine wheel. Now the state of the art small GT25 wheel used flows as much as a fairly large trim but old school T3. The GT25, with its low inertia, conventional wisdom says it should spool very quickly. However with the large size differential between the compressor and turbine, the large slip losses make this turbo a laggy unresponsive dog despite the low inertia, high flow and the ball bearing center section that this particular turbo has.

Here the differences between a 84 trim (left) and a 76 trim turbine wheel are shown. Note the exducer, diameter in relationship to the inducer on the two wheels.

The B-Width (the area indicated by the arrows) on the turbine wheel on the left is much smaller than the turbine on the right. Bigger B-Width usually = bigger flow, all other things being equal.

2. Although I have no math to prove it, I feel that exducer diameter can be a pretty good way to estimate what turbine is appropriate for the expected power output within turbine families. Although a turbine's power-supporting capability depends on a number of other factors such as; turbine trim, B-Width, blade number and blade shape and housing A/R, exducer diameter is the easiest thing to fathom when "sizing up" a turbine. To explain some of these terms, turbine trim is the same mathematical ratio as compressor trim, B-Width is the blade tip height on the turbine wheel inducer. Typically, the taller the B-Width, the more flow a turbine has. The blade sweep and blade shape has a big effect on performance that is not possible to just size up by eyeball, the blades are swept to reduce what us engineers call stagnation losses, a whole subject in itself. The more blades a wheel has and the thicker the blade, the less a wheel will flow. Blade thickness can affect turbine efficiency also because they are less efficient. Housing A/R is explained in detail shortly. Since there is no empirical data on turbines available to the public from the aftermarket, my exducer size method of turbine sizing is basically an educated-guess method based on lots of screwing around with turbos on my part. I don’t claim that this is the best way to figure things but on the information available to the general public, it is better than not considering things with any sort of logic! Noted

Innovative Turbo Systems' radical R trim turbine has a trim of 100. Note how the inducer and exducer are the same diameter. This is for maximum flow. Don't even think of using this on your Honda.

TURBINE SIZE GUIDE

Turbo Family	Exducer Bore (inches)	Trim with Housing A/R	Est. Power Level
T25/T250/T28	1.642	62	140-170 hp
	1.855	79	200-320 hp
	1.918	84	250-350 hp
T3/T31/T350	2.122	69	200-300 hp
	2.229	76	300-400 hp
	2.439	76	400-500 hp
TO4*	2.544	76	500-650 hp
TO4**	2.693	75	550-750 hp

* Not too streetable, some leading Quick Class racers run this wheel

** Probably too big for any current 4-cylinder Honda motor

turbo guru Corky Bell in his book *Maximum Boost* also gives exducer diameter as a rough starting point for turbine power supporting estimation. Real turbo engineers are going to gnash their teeth at this hack, but I challenge them to try this on an actual car—they will find it works!

3. Smaller trims of the same wheel family will spool faster but flow less well. The larger variance inducer to exducer gives the exhaust gas more oomph! on the wheel. It also gives the gas a less direct path out of the turbine, giving more backpressure.

In the chart above I offer some of my estimations based on what many people are running both turbowise and powerwise with some popular, readily available Garrett applications on 4-cylinder engines. These are some examples available from Turbonetics and Innovative Turbo Systems.

Defining A/R and How to Choose The Right Exhaust Housing

You have probably heard the term *A/R* used in some context or another in this article or when talking to turbo people about turbos. A/R is one of the most critical aspects in turbo tuning. Selecting the proper A/R for the turbine housing is one of the major things a tuner can do to control how the turbo responds to throttle input. The turbine housing is the cast iron casting that contains the turbine wheels. Adjusting the housing A/R can control much of the turbo's operating characteristics.

A/R stands for area ratio and is a mathematical description of the cross sectional area of the turbine housing inlet divided by the radius described by the center of the turbine wheel to the center of the turbine housing inlet. This ratio stays constant throughout the volute (snail-looking part) of the housing as it tapers from the tight part of the snail shell to the turbine-housing inlet.

A large A/R housing has good flow, less backpressure and creates more top-end power at the same boost pressure by improving the volumetric efficiency. It will also spool slower. Conversely, a small A/R housing has

Externally there is not much difference between a large 0.82 A/R (left) and a smaller 0.63 A/R T-3 exhaust housing (right) even through they perform very differently.

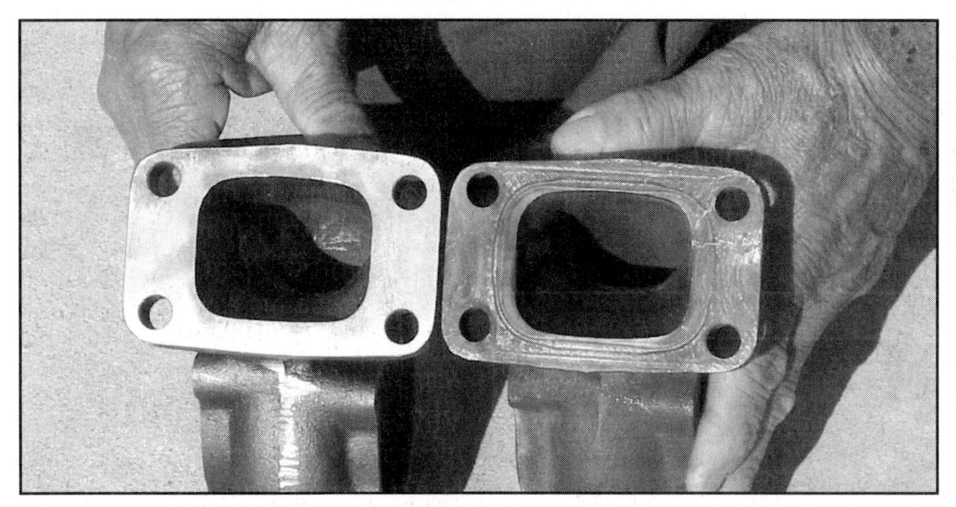

The view inside the housings is much different, the 0.82 A/R housing (left) has a much bigger internal volume so it flows better, this gives less backpressure and more power. The 0.63 A/R housing (right) has a smaller volume which acts like adding a nozzle on a water hose, speeding spool up at the expense of backpressure and top end power.

Lisa Kubo's pulse converter manifold enables the use of this huge T51 turbo without excess lag. She is currently the world's fastest unibody Quick Class FWD Honda.

less flow and more backpressure but has quicker turbo spool-up characteristics.

If you are wondering why this is so, consider this: Imagine spraying water at a pinwheel with a garden hose. When you simply spray the water on the pinwheel, it spins up to speed slowly. The water flow out of the hose is good but the pinwheel takes longer to spin up to speed. Now put a garden nozzle on your hose. The thin, high velocity stream of water spins the pinwheel up rapidly. There is more backpressure in the hose and the overall water flow from the hose is much less but the pinwheel really gets up to speed fast. This is an extreme example but is exactly how A/R works inside the turbo.

Following are some commonly available A/R turbine housings for various Garrett turbo families. These are different sizes stocked by Turbonetics and Innovative Turbo Systems:

T25/T250/T28: 0.48*, 0.63, 0.64, 0.86.

T3/T31/T350: 0.36*, 0.48, 0.63, 0.82

TO4: .50, 0.58, 0.70, 0.81**, 0.84**, 0.96**, 1.00**, 1.15**

*Probably too small for any Honda applications.

**Probably way too big for most Honda applications.

Turbine Housing Configurations—There are basically two types of turbine housing configurations, tangential and on-center. Tangential housings have the turbine inlet coming into the housing at a tangent to the turbine wheels. On-center housings have the turbine inlet coming directly into the center of the turbine wheel's axis. A tangential turbine housing is greatly more efficient as the gas path feeding the turbine has a straighter shot into the housing and a better impingement angle on the wheel. The gas entering an on-center housing must go around a kink in the housing, which causes turbine efficiency to suffer by up to 10%. There are studies that show that a 2% gain in turbine efficiency can offset gains of up to 25% of the turbine's inertia so a 10% loss in efficiency is quite significant. However, the on-center mounting can often give a bit of mounting flexibility to solve a difficult fit problem but because of the efficiency losses it is better to try to find almost any other solution to a packaging problem.

Another interesting aspect of turbine housing design is divided and undivided housings. A divided housing is exactly how it sounds: the scroll of the turbine housing is split in two. A savvy tuner can use a divided housing to his advantage on an engine with few numbers of cylinders. A divided housing works best on a 4-cylinder engine with some advantages on a 6-cylinder with a properly designed manifold. When a divided housing is used, usually cylinders 1 and 4 are fed into one side of the scroll and cylinders 2 and 3 are fed into the other side. The cylinders fed into each side of the scroll are as far apart in the firing order as possible. This allows the turbine to be hit with 4 distinct pulses as the engine goes through its firing order. This improves turbine efficiency, sometimes to the point where up to one size larger A/R housing with its attendant lower backpressure can be used, either that or less turbo lag can be enjoyed with the same size A/R housing. The divided housing can also improve volumetric efficiency by making reversion from adjacent in firing order cylinders much more difficult. This is because there is a great deal of separation in degrees of crankshaft rotation between the valve opening events of the adjacent cylinders. In order for a reversion pulse to contaminate an adjacent firing cylinder, it has to travel back through the spinning turbine blades and up the other side of the divided turbine housing scroll to get to the adjacent cylinder. This is pretty difficult and the pulse will tend to take the path of least resistance, past the turbine to the area of lower pressure.

A divided exhaust turbine housing is used on a tube framed Outlaw Class racer to help spool its monstrous T61 compressor.

On more than 6 cylinders, the divided housing is probably not worth the effort except perhaps in a V-8 engine with a 180-degree crank in twin turbo configuration. In this case, the V-8 can be treated as two 4-cylinder engines.

Lisa Kubo, Steff Papadakis and Myles Bautista have examples of manifolds designed to take advantage of a divided exhaust housing. Check them out in the pits at the next race.

It is probably best to avoid extremes of A/R selection as turbine efficiency falls off greatly when selecting a very small or very large A/R turbine housing. If you feel the need to go to a very small or very large A/R, then perhaps you picked too big or too small of a turbo combination for what you really wanted to do. You need to remember that you must be realistic when picking a turbo size. A 600 hp racing monster turbo is not going to be a quick spooling lively street turbo no matter what A/R housing you put on it.

Some interesting things to note is that it is often better to put a small trim turbine wheel in a bigger A/R housing to solve a spool-up time problem rather than putting an extremely small A/R housing on a large trim turbine. The larger A/R housing can offer more expansion across the turbine allowing more energy recovery by the turbine to help spool the turbo up quicker. A good way to think of this is to try a smaller turbine trim but in the same housing, not to go to extremes. Turbines usually work the most efficiently in housings around the mid-range offerings in A/R sizing. Generally the turbine/housing combo was designed around this mid-size and the blade angles, etc are optimized for this, another reason to avoid extremes either way in A/R when configuring your turbo.

Interestingly enough, compressor housings are also available in different A/R combinations but compressor performance is not linked as closely with housing A/R as it is on the turbine. Smaller A/R compressor housings tend to be slightly more efficient at higher-pressure ratios than bigger A/Rs with bigger A/Rs being slightly better at lower pressure ratios. Bigger A/R housings also tend to come on slightly later in the power curve and more violently.

So in short, when sizing up a recommended turbo combination for your ride, you pick your compressor size by the included calculations and the compressor map, then pick the applicable turbine for your power requirements and fine-tune the powerband of your configuration by the A/R of your turbine housing. Be realistic: don't pick a drag racer–sized turbo if you are seldom going to hit the track.

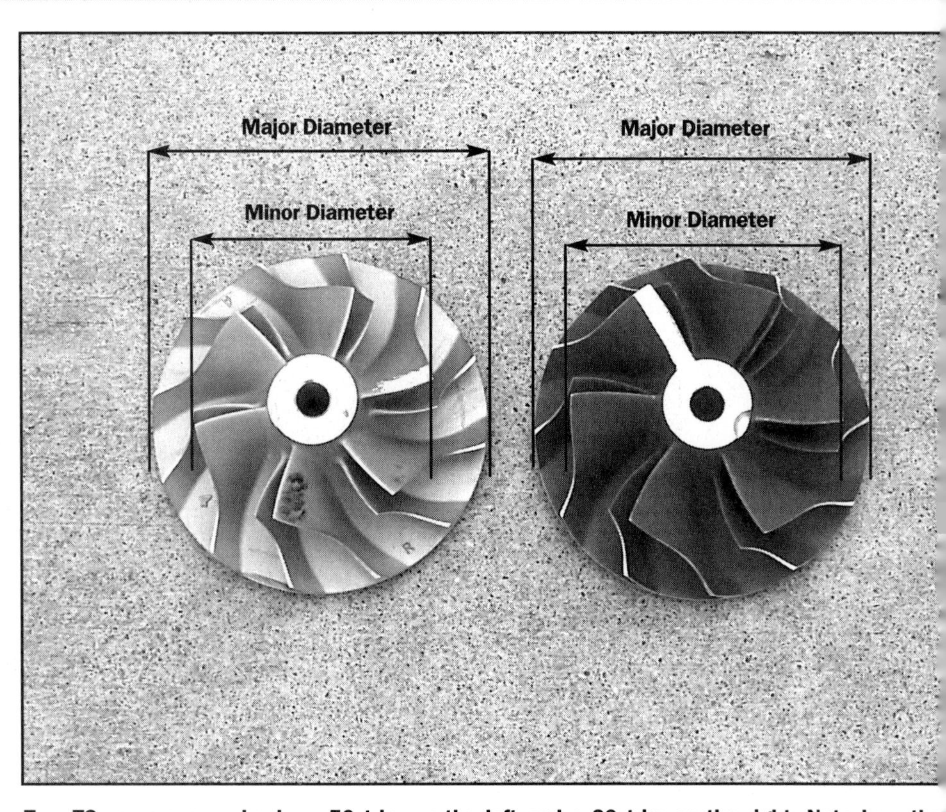

Two T3 compressor wheels, a 50 trim on the left and a 60 trim on the right. Note how the inducer of the 60 trim is larger than the 50 trim. This allows the 60 trim to have more flow.

What Is Trim?

When talking to turbo people, you will frequently hear the term *trim* thrown around. What is it? Trim is a term to roughly describe the size of a turbocharger compressor or turbine wheel within the family of turbo. The term *trim* is used rather loosely in the turbo industry. It could be a commercial marketing term like Turbonetics uses to describe many of their wheels, like O, P and Q for their TO4 family of turbines with the O being the smallest trim and Q the largest. Conversely Turbonetics uses S, V and H to describe the trims of their TO4 compressors. It can also be the actual mathematical definition of the wheel size, like the Garrett TO4E compressor in 46 trim, 50 trim, 57 trim and 60 trim. Garrett also uses 62, 68, 76, 79 and 84 trim etc. for their turbine wheels.

What is the mathematical formula for trim? Figuring out the mathematical number for it is really quite simple:

Trim = (minor wheel diameter / major wheel diameter) squared x 100

This is the relationship between the major and minor diameters of the wheel or the total diameter and the inducer diameter (for the compressor) or the exducer diameter (for the turbine). What do these numbers mean? Generally the bigger the trim number, the more flow the wheel will have.

Bigger trim compressor wheels tend to be a few points less efficient than the smaller trim wheels because of the mach differential across the face of the inducer blades, especially with bigger compressor wheels. The face speed of the blade is higher at the outer ends of the inducer than in the inner edge because of the distance traveled by the blade every revolution. This speed differential somewhat screws up the efficiency and is one of the reasons why the blades are swept and contoured like they are in a well-engineered compressor wheel. The other reason is to reduce stagnation losses but this is getting way beyond the scope of this book.

Usually it is a better compromise to use a bigger trim smaller wheel and give up a couple of points of efficiency, than use a bigger wheel and gain more lag-causing inertia.

In turbines, a bigger trim in the same wheel size family will flow better and have less backpressure but recover less energy from the exhaust flow and take longer to spool than a smaller trim wheel.

How to Calculate Intercooler Efficiency

How well is your intercooler working? Do you need to upgrade it? It is pretty easy to calculate intercooler efficiency. All that is needed is a digital thermometer and time to make a few full-throttle passes. Many digital multimeters have a function where a thermocouple can be plugged in to take air temperature measurements. Digital thermometers can be purchased relatively cheaply. All one has to do is measure the air temp before and after the intercooler during a full throttle third-gear pull. One should try to be consistent and do the pre-intercooler and post-intercooler pulls to at least past the engine's torque peak, staying on the throttle for the same amount of time. After collecting the temperature data, the data can be fed into this formula to get intercooler efficiency.

E=T1 – T2/T1 – TA
Where:
E = Efficiency in percent
T1 = Compressor outlet temperature
T2 = Intercooler outlet temperature
TA = Ambient Temperature

As an example, let's say your car has a compressor discharge temp of 250 degrees; a post intercooler air temp of 100 degrees and it is a 70-degree day.

E = 250-100/250-70
E = 83.3%

As a guideline, factory turbo car intercoolers typically have intercooler efficiencies as low as 50% while a good drag racing ice water intercooler can be over 90% efficient. A big air-to-air intercooler with good airflow can be over 80% efficient.

Lisa Kubo's fuel system has a Paxton high-capacity regulator, a big fuel rail and big RC Engineering injectors to feed her thirsty 650+ hp motor.

Calculating Fuel, Injector Size and Fuel Pump Capacity

It's a fact; turbos need lots of fuel to make the big power numbers. It is critical that your fuel system can supply the extra fuel needed to fuel the extra air that the turbo is pushing in. The math needed to figure out what size injectors you will need for your turbo motor is pretty easy, nothing like the gyrations for compressor mapping!

Injector Size—Injectors are rated in either lbs. of fuel/hour or cc/min. American injectors, like MSD, are usually rated by lbs/hr while the injectors sold by Japanese tuning companies are usually rated in cc/min. Here is how we figure out injector sizing:

Injector size in lbs/hr per injector = Expected hp x BSFC/ Number of injectors

For BSFC, figure on a conservative 0.60 for a rich running, pump-gas-burning turbo car where the tuning will be on the rich side to prevent detonation, or if you simply want to be conservative, 0.50 is better if you are going to have a crisply tuned race car on race gas. 0.45 is maximum tuning right on the edge of destruction. It is probably better to use 0.50 anyway so you have a little extra fuel latitude. To convert to cc/min:

Injector size in cc/min = [(Injector size in lb/hr x 454)/60]/ 0.75

Fuel Pump Capacity—Likewise the math needed to figure out what size fuel pump you will need is pretty easy. American fuel pumps are usually rated in gallons per hour. Import fuel pumps are usually rated in liters per hour. It is important to make sure that your pump gives its rated flow at high pressures. Fuel injected engines run 35 psi of fuel pressure and higher, unlike carburetors that run at less than 10 psi. Make sure your pump is a performance high-pressure pump for fuel injection. To figure out what size fuel pump will be needed:

Total fuel flow = injector fuel flow in lb/hr x number of injectors
Fuel flow in gallons per hour = Total fuel flow in lb/hr/6.25

To convert to liters/hr:

Fuel flow in gallons per hour x 3.8= fuel flow in liters per hour

TURBO UPGRADES

If you have a turbocharged Honda and want more power from it you often can upgrade the turbo that came with your turbo kit. The correct way

This is what the diffuser with a large T3 housing with a T3 wheel looks like. Note the wide diffuser.

When a big TO4 compressor wheel is put into a smaller T3 compressor housing, the diffuser area gets way too small. This has a greatly negative effect on compressor efficiency. This amount of diffuser reduction is seen in some poorly engineered aftermarket upgrade turbos. Let the buyer beware!

A clipped turbine is not a good way to gain more turbine flow; it is only done as a last resort to try to tweak some flow out of a turbine or as a cost savings measure. This turbine is clipped about 20 degrees, the maximum a turbine should be clipped.

to do this is to use a larger trim compressor and exhaust wheel with the proper housings on your existing turbo. In the T3 and TO4 Garrett turbo families there are many choices when it comes to different wheels and housings. A proper way to upgrade would be to get a larger trim compressor wheel from the same family of turbo and to go up one size in exhaust housing A/R on the turbine side.

A common no no that is often done is when the compressor housing is machined out for a very large diameter compressor wheel from a larger family of turbo. This is bad because it can reduce the diffuser in the compressor housing to almost nothing. The diffuser is the parallel walled area of the compressor housing just past the outer edge of the compressor wheel. This is where much of the compression of the air takes place. If the compressor wheel takes up most of the diffuser, the compressor becomes what is called a Scroll diffuser compressor. This is like how the very first turbos were designed, which are a lot less efficient than modern turbos. If an upgrade has significantly cut down the diffuser width, you can expect compressor efficiency to fall to around 50% in this case or about the equivalent of a really bad Roots blower!

An inefficient compressor requires a lot of turbine power to produce boost, which contributes to backpressure and lag. A compressor upgraded like this may flow more and make more power but it will pump hot air and contribute significantly to turbo lag and a possible reduction of VE. Better hope your intercooler is very efficient! The reduced efficiency means that the charge air temp can be over 100 degrees F hotter than a good compressor.

A proper turbo upgrade leaves most of the diffuser in place by using a compressor wheel of the same type and family but of a bigger trim or by using the correct matching housing for the compressor wheel used.

Clipping the Turbine Wheel

A method often used in upgrading the turbine side of things is clipping the wheel in an attempt to gain more flow. Clipping is grinding the turbine blades at an angle usually 20–30 degrees from the root of the turbine wheel exducer to the blade tips of the wheel exducer. This takes some of the bend out of the turbine blades, enlarging the throat of the turbine wheel much like a larger trim does but not as well. This supposedly gives the exhaust gasses a straighter shot going out of the turbine.

What this really does is kill turbine efficiency. Surprisingly we have seen where it sometimes causes more backpressure by throwing off the compressor-turbine power balance. At the least it contributes to lag. A bigger trim turbine wheel and/or housing A/R are the preferable way to go here instead of a clip. If a much bigger wheel is to be used, it is better to get a matching housing designed for the turbine. However the turbine is not as sensitive to the housing geometry as the compressor is so you have more leeway when carving out a stock turbine housing for a bigger turbine wheel. When examining a carved for a larger turbine wheel turbine housing, make sure that it is machined correctly to accommodate the larger turbine wheels usually wider B-width. A turbo upgrade that has a lot bigger of a compressor wheel than turbine wheel should be avoided as slip losses also kill efficiency, create backpressure and create more turbo lag.

Single or Twin Turbo?

Although it is a beautiful technological terror, the Apexi drag race Integra's twin turbo system probably gives up a little in both response and power to a well-engineered single turbo system.

This is somewhat controversial but many people believe that twin turbos spool faster than a large single turbo. This is because twin turbos use two smaller turbos with less inertia than one bigger single turbo. This is somewhat true for larger engines like the bigger six cylinders and V configuration motors. The twin turbo will be slightly better in transient response than a bigger single turbo but not a whole lot.

Smaller turbos tend to be less efficient than their bigger brothers because of the scaling of their wheel to housing clearances. The bigger turbos only have slightly larger clearances between the housing and the wheels than their much smaller little brothers. This clearance causes internal leakage, which reduces efficiency of both the compressor and turbine. Since the clearances and the leakage are proportionally larger on the smaller turbos, they are less efficient, and thus laggier than you would think they should be. There have been published studies showing that a 2% gain in turbine efficiency can overcome a 25% gain in the turbos rotating inertia so even small efficiency gains can be significant. The ratio of blade thickness to wheel size does not scale up directly either. Smaller turbos often have proportionally thicker blades on their wheels for their size than larger turbos; thicker blades are aerodynamically less efficient. Because of these issues, for 4-cylinders, a big single is the only way to go, especially when the large pulses from a 4-cylinder can be used to spool the turbo faster in a well designed system that exploits a divided turbine housing.

The only time a twin turbo should be seriously considered is on V-6 and V-8 motors, where it packages well and you can use two larger turbos. On 4-cylinders, consider a twin setup only for show purposes.

It is possible for poorly upgraded turbos to have more power than stock but it is a brute force way of getting things done and most of the time the added power will be in a narrow, high band. Much more power with less stress on the engine can be had with a proper upgrade using bigger matched wheels and housings from the same family. Correct fueling, intercooler efficiency and spark management is more critical with a backpressure bound, inefficient, poorly upgraded turbo.

Extrude Honing

One upgrade that works really well is the Extrude Hone abrasive polishing process, especially on the exhaust turbine side of the turbo. We have experienced that a turbo given the Extrude Hone treatment can spool over 500 rpm sooner and flow better. Lisa Kubo's car experienced a 50 hp gain across the board when her turbo was given Extrude Hone. Several other top racers have also experienced similar gains in power after treatment. As the process is a little on the expensive side, if you can only afford some of the process, do it to the exhaust housing. Testing has proven that this is the largest gain of any

Extrude Honing the wheels thins the blades slightly and gives the wheels a smooth polished finish. Both of these factors improve the wheels aerodynamic efficiency.

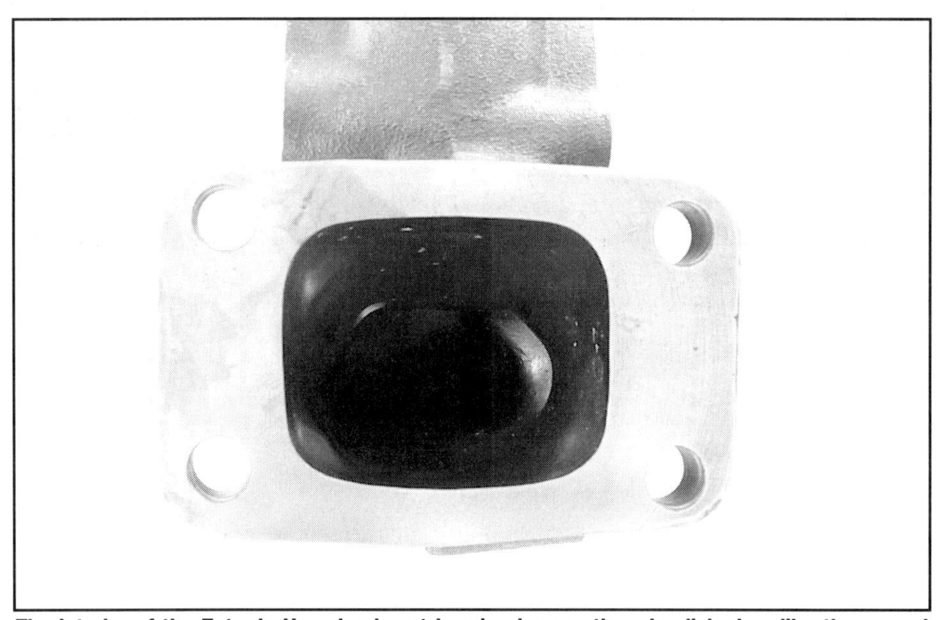

The interior of the Extrude Honed exhaust housing is smooth and polished, unlike the normal rough cast iron it normally is. Extrude honing the exhaust housing probably makes the biggest difference of all the turbo parts.

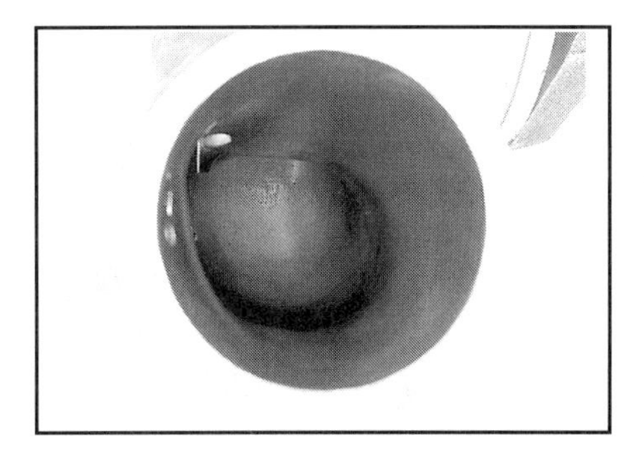

This is a close-up of the interior of an Extrude Honed compressor housing. If you are on a budget and can only afford to do one side of the turbo, the exhaust housing makes a bigger difference.

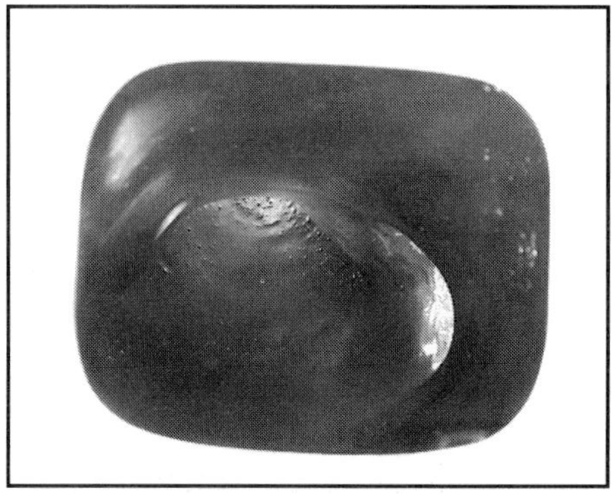

Here is a closer view of the smooth interior of an Extrude Honed exhaust housing. Doing the exhaust housing makes the biggest difference of all.

single turbo part to treat.

Porting

Porting the lumps and bumps out of the turbine housing near the turbine housing inlet is a cheap, easy-to-do and effective upgrade you can do yourself. If you are going to Extrude Hone your exhaust housing, do this before sending your turbine housing to Extrude Hone.

To sum up, an upgraded turbo can work well or it can be a real engine-stuffing lag monster at best or at worst, a lag monster with less power than stock. It all depends on how the upgrade was done.

TURBO COMBINATIONS

Did all this math scare the hell out of you? Do you want to blindly trust us? Well, if so, there are some combinations in the chart nearby that should work fairly well with different sized motors. When we are talking motors we mean 4 valves per cylinder, 4-cylinder Honda motors. These recommended turbos are commonly available in the aftermarket Garrett parts. Why Garrett? Garrett is the world's largest manufacturer of turbochargers and they make a plethora of different wheel, housing sizes and trims to closely match nearly every conceivable engine combination. Garrett compressor maps are also available in many aftermarket companies' catalogs or available upon request so you can do some of your own turbo engineering when armed with this book. No other turbo company has this information as available to the general public as Garrett.

The matches here are close to the maximum for a streetable car but will give the flexibility to make quite a bit of power if race gas, different tuning and more boost is used. They will have some lag but a manageable amount, some flexibility in tuning with different exhaust housings and best of all, will outpower any available supercharger kit when applicable. These combinations are available inexpensively through aftermarket Garrett distributors like Turbonetics and Innovative Turbo Systems. These combinations might not be what the turbo salesman will recommend and are certainly not the only possible ones that will work well out there but they should work well if your end use is as what we describe in the chart below.

MAXIMUM STREET PERFORMANCE COMBINATIONS

Engine Size	Compressor Trim/ Housing A/R	Turbine	Exhaust Housing A/R	Strong Boost Onset rpm	Approx. Power Potential
1500-1800cc	T3, 60-62 trim	T25, 79 trim 0.60 housing	T25:0.63-0.86 T3, 69 trim	3000-4000 T3: 0.63-0.82	180-300
B16A* 1600cc	TO4E, 46 trim	T31, 76 trim 0.50 housing	T3: 0.48-0.63	4500-5500	220-350
1800-2200cc	TO4E, 50 trim	T31, 76 trim 0.50 housing	T3: 0.63-0.82 T350, 76 trim	3500-4500	300-420
2200-2500cc	TO4E, 50-60 trim	T31, 76 trim 0.60 housing	T3: 0.82 T350, 76 trim	3500-4000	300-500

Note: Of course the boost will actually go positive 1000-1500 rpm sooner than the rpm numbers given, but this rpm shown will be where all hell breaks loose and a big fat surge of power will be felt. We feel that these combos are very streetable but we included this data so you can decide whether or not you think it's streetable. If you are unsure, err to the smaller side of housings and wheel trims. Remember it is often better to go to a smaller trim turbine in a bigger housing or a smaller trim in the same housing than vice versa to get better spool.

As for horsepower numbers, this depends on the exhaust housing used, the octane of the fuel used, the boost pressure run and to a large degree, the skill of the tuning involved. The higher end of the power scale we give represents tuning on race gas.

* Due to the high revving capability of the B16A, it can handle a bigger turbo if the power is desired and lag can be tolerated.

For those of you who are unmotivated or just need a fast spooling combo for autocross, pure street driving or simply don't want much power and want power at low rpm, the following combos will work good for fast spooling and low boost levels (10 psi and below). They will also approximate the power output and throttle response of a good supercharger kit.

QUICK SPOOL/LOW BOOST COMBINATIONS

Engine Size	Compressor Trim/ Housing A/R	Turbine	Exhaust Housing A/R	Strong Boost Onset rpm	Approx. Power Potential
1500-1800cc	T3, 50-60 trim 0.48-0.60 housing	T25, 68-79	T25:0.63-0.86	2000-3000	150-250
1800-2200cc	T3, 60-62 trim 0.60 housing	T25, 79 trim	T25: 0.86	2000-3000	200-300
2200-2500cc	TO4E, 46-50 trim	T31, 76 trim	T3:0.63	2400-3000	250-350

Finally for the racers out there, here are some all-out, balls-to-the-wall combinations to go after the 10 and 9 second ET's. These combos require a seriously built motor and race gas only to support the massive airflow that they are capable of. They will be soggy and unresponsive for street driving. High RPM capable motors can go to the next higher turbo size recommendation for maximum power and additional lag.

FULL RACE COMBINATIONS

Engine Size	Compressor Trim/ Housing A/R	Turbine	Exhaust Housing A/R	Strong Boost Onset rpm	Approx. Power Potential
1500-1800cc	TO4E, 50-60 trim 0.60 housing	T31, 76 trim	T3: 0.82	5000-6000	350-450
1800-2200cc	TO4E, 60 trim	T350, 76 trim 0.60 housing	T3: 0.82	4500-6000	400-500
2200-2500cc	T61, 46 trim 0.50-0.60 housing	TO4, 76 trim	TO4: 0.70*	4500-6000	500-650

*Divided housing

ALL-OUT ALL MOTORS 10

Maybe a turbo is not for you. Maybe you want to race in a class with different challenges. Or, maybe you are studying ways to get the most out of your naturally aspirated motor. Perhaps the class you are road racing in does not allow turbos but allows extensive engine modifications. Perhaps you want to study what the drag racers are doing for your road race motor. Could it be that you have a pathological hatred of forced and chemical induction? If so, this chapter on the latest in all-out naturally aspirated Honda motors could be exactly what you were waiting for.

It's no secret that the formula to build a fast turbo Honda has pretty much stabilized. The technology for turbo motors has matured and gains are starting to become incremental. Just about any import event will have all of the Quick Class Qualifiers in the 10s and at least 3–4 in the 9s. You can now pick up the phone and order yourself a 9-second Honda, just as if you were ordering an expensive pizza.

In import drag racing there is a new class heating up where innovation is ripe and the technology is changing quickly. It is the naturally aspirated all-motor class where the racing is most interesting and records are being swapped on a race-by-race basis. This is an engine builder's paradise: it is an engine-only, no power adder, mano to mano challenge to extract power from thin breathable air. Because the crutch of the boost controller does not exist here, every bit of power helps and the tuners strive to eek out every advantage possible.

The world's fastest naturally aspirated Honda at home in the pits.

Unbelievably these all-motor cars are getting consistently into the 11s and a few are unbelievably dipping into the 10s where they could, on some weekends, qualify mid-pack for the turbocharged Quick Class!

How do these cars do it? What is inside these motors? What does the future hold? We are going to look at the insides of a few of the class leading cars, which is secret stuff. Many lives were lost in this endeavor. We had to do some late-night sneaking around with our digital miniature spy camera, buy drinks to loosen tongues and other devious activities to get these secrets, but get them we did, all for the sake of getting the inside scoop.

First let's get right to the nuts and bolts of the matter. Just what is inside these high revving, speed-shifting, high-strung monsters? To find out, we'll look at two motors from two of the leading, naturally aspirated engine builders on the import drag racing circuit today. These two builders both have fast, record-holding and race-winning machines. Both of these builders have methods of building motors that are diametrically opposed. Both engines are built with dramatically different theories and viewpoints. Both are damn fast and fierce competitors. Strangely enough, both engines seem to match the personalities of their creators.

A view of the twin TWM throttle bodies in Jeremy's car brings back memories of side draft Weber carburetors.

R&D RACING

Let's first look at the motors of R&D Racing. R&D is a hotbed of naturally aspirated (NA for short), Honda motor development that currently holds the record for the fastest all-motor run to date, a blistering 10.68 at 129.8 mph owned by Jeremy Lookofsky.

R&D is also the leader in SOHC D motor engine development, building Bisi Ezerioha's fastest all-motor SOHC CRX, which has run a best time of 11.40 at 121 mph as this book goes to print. As the well-known B motors little brother, the D never gets much respect from the performance enthusiast, but not this one.

What is in R&D's secret motors and what makes them tick? Read on and find out.

Let's look into the guts of Jeremy Lookofsky's and Bisi Ezerioha's record-winning motors. Darren SanAngelo, R&D's owner, is the brains behind these fast creations. SanAngelo's motors adhere to the "no replacement for displacement" school of thought. They are big in displacement. Unbelievably big, so big that if some of the engineers at Honda could see what was going on with their marvelously designed motors, they would probably say that it was either impossible or flat die of a cardiac arrest on the spot.

R&D's racing motors also push the limits of thermal efficiency, fuel science and combustion stability with sky-high compression ratios, more than we have ever heard of being successfully run in any sort of racing before. The result of this approach is motors that are tractable with lots of torque and broad useable powerbands, sort of like the quiet, soft-spoken SanAngelo's own low-keyed personality. These characteristics also are passed onto R&D's awesome detuned NA street motors that deliver supercharged hp with NA reliability.

THE H22

Block

Let's look inside Jeremy Lookofsky's Honda. Starting with the block, we see that R&D has used a third-generation Prelude H22. This variant of the H22 was selected because it features a much stronger closed deck. The later fourth-generation blocks have a weaker, more flexible, open deck which must have more extensive tweaks, mainly a welded-in block guard or expensive Darton deck reinforcing sleeves to withstand the loads that racing places on it.

The stronger, third-gen Prelude closed deck block is fitted with R&D's ductile iron sleeves, which allows a big oversize piston to be used. Although the stroke, bore, rod length and displacement are secret, Jeremy's motor packs a whopping 2600cc of displacement. That is more than 400cc's over stock! The bore surfaces are finely finished so the low tension rings can seat and seal quickly. Darren disconnects the balance shafts in the block on race motors. This gains as much as 10 hp due to friction reduction. Darren leaves the balance shafts in for street and endurance type motors for improved oil pump life.

All of the main cap and head bolts are replaced with ARP studs to maintain a higher clamp load for better head gasket sealing and for additional strength for the bottom end. The block is fully deburred, all press-in oil passages plugs replaced with screw-in and the block is notched at the base of the cylinder bores for rod

Here are the reciprocating parts of Jeremy's big 2600cc H22 engine. A forged stoker crank, billet rods and forged pistons with low-tension rings and tool steel pins. Although this high dome, high compression piston is for race fuel only, with a lower compression ratio, this would be one hell of a street sleeper motor.

clearance needed with the big stroke.

Crank

R&D has their own crank blanks made from 4340 forgings. 4340 is a tough high-nickel chrome-moly alloy that is considerably stronger than the stock steel crank. The unique forging is pretty cool as it allows better alignment of the grain structure within the crank, assuring that it will be plenty strong. R&D makes the crank blank with a lot of extra material around the rod journals to allow for generous stroking of the crank during final finish machining, if so desired.

Rods

R&D's own rods are used here, available in either H or I beam pattern. The rods are cut from 4340 forgings and feature 220,000-psi H11 tool steel rod bolts. Like the crank, R&D has commissioned that their forging dies be made to their spec, so these rods are lighter than your typical aftermarket rod. The prices for the R&D rods are pretty reasonable when compared to typical racing custom billet pieces.

Pistons

R&D low-silicone, pistons are hand massaged to an incredible, unheard of, 16:1 compression ratio (estimated), requiring the use of very high octane racing gas! Conventional wisdom says that compression ratios of above 14:1 are in the zone of diminishing returns, as combustion stability becomes an issue at this point. Somehow the guys at R&D have figured a way to keep the fire lit with this hard squeeze. The pistons are aggressively gas ported in an effort to maintain ring seal at astronomical piston speeds.

The forged pistons are fitted with Speed Pro piston rings with the piston being gas ported for better high rpm ring seal. The gas porting and lightweight low inertia, low-tension rings are important to maintain good ring seal and avoiding ring flutter at the engine's long stroke, short rod

Large cuts need to be done at the bottom of the cylinders to clear the rod on Jeremy's stroker H22.

The combustion chamber of the R&D H22 shows much attention to detail. Stainless steel, 1mm oversize valves are set in the polished unshrouded chamber.

AEM adjustable timing gears spin secret grind R&D camshafts set on unspecified lobe centers.

induced ultra fast piston speeds. The pistons also feature a gas trap between the number one and number two piston ring to further improve ring seal. Lightweight full-floating H11 tool steel-tapered pins hold the pistons to the rods.

When questioned about piston speed issues, Darren stated that the engines still produce good top-end power. This says loads about Honda's excellent flowing ports and says volumes about the headwork.

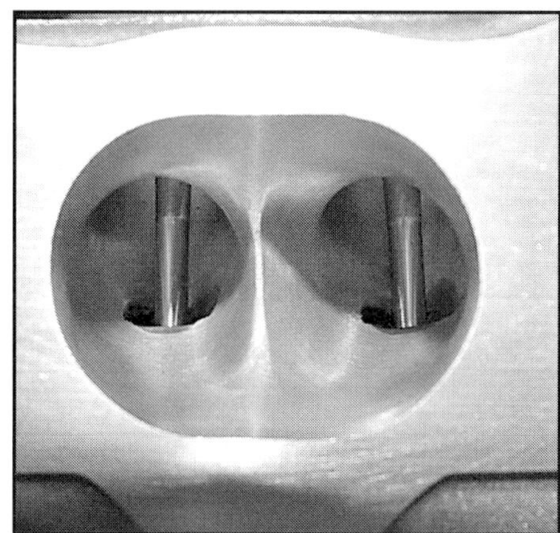

R&D's H22 exhaust port uses conventional porting techniques. Just good normal porting to the line of sight.

The Jeremy's H22 valvetrain is kept under control with Ferrera dual valve springs and titanium retainers.

Although Darren isn't talking about these specs, we figure that these motors by nature have to run pretty low rod-to-stroke ratios, which makes port flow even more critical, especially at high rpm.

Cylinder Head

The R&D cylinder head is pretty straightforward, with slightly larger, tuliped and undercut stem stainless steel valves with stock valve guides with clean, straightforward line of

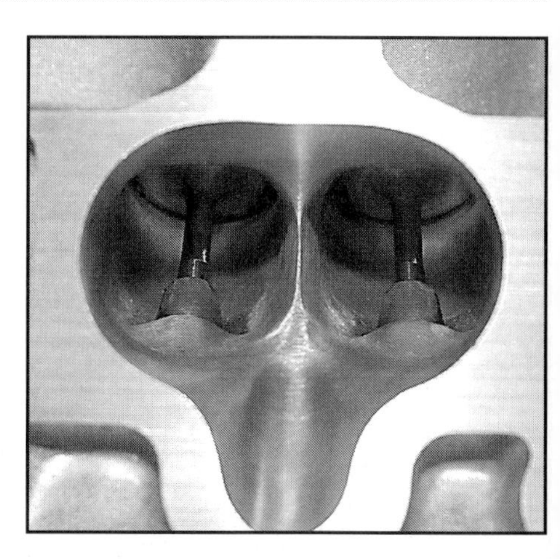

The intake port on Jeremy's H22 head shows line of sight porting and good workmanship.

sight porting. A narrow but otherwise conventional 45-degree seat, three-angle valve job is used. The combustion chamber is cc'ed and polished with minor unshrouding around the valves, nothing crazy here, just good straightforward work.

Valvetrain

The heads are fitted with double valve springs and titanium retainers. Although the exact specs are secret, the R&D cam has around 0.500" lift and about 310 degrees duration. The cams are timed with adjustable timing gears to a secret unspecified lobe centerline. These cams produce power to the engine's 8000 rpm redline.

The Top End

The engine is fitted with R&D's custom in-house design, stainless steel, 4-1 headers which feature a stepped primary diameter and a merged collector. A TWM manifold handles induction chores with dual 48mm sidedraft throttle bodies fired by a Speed Pro engine management system. The spark is provided by an MSD ignition system. The head is given straightforward porting and makes use of sidedraft throttle bodies driven by a Speed Pro engine

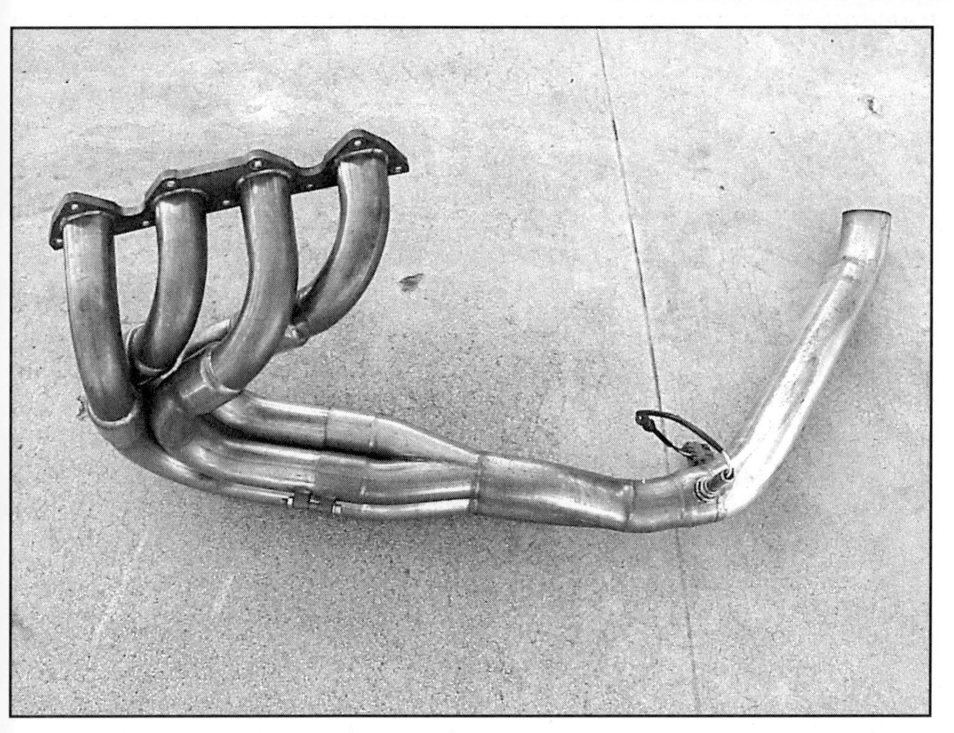

The radical header used on Jeremy Lookofsky's car features stepped diameters and anti-reversion chambers, which are the bulges on the primary tube.

Jeremy's car uses twin 48 mm TWM throttle bodies on a TWM manifold to handle fueling duties. The throttle bodies provide a sharper throttle response and are an updated offshoot of old-school dual side draft carburetors.

management system. Although it is usually shifted at 8000 rpm, the big motor can be spun to an impressive 8800 rpm.

How much power do all of these parts buy? Jeremy's record-setting motor pumps out approximately 260 hp @ 7800 rpm to the wheels with a whopping 200 lbs-ft. of torque! The powerband is broad, wide and flat with an equally flat torque curve, not your typical Honda high rpm, low torque peaky powerband at all. Surprisingly, this monster motor still likes to rev, pulling strongly with no drop-off in power to the redline.

Darren mentioned to us a cool point; that this motor can be detuned for pump gas by lowering the compression ratio to 11:1 and installing a milder cam. Tuned like this, the motor will still put out 220-wheel hp and 180 lbs-ft. of torque. This is about that of a supercharger kit but with a broader powerband and NA reliability. That street motor would be a real sleeper. With VTEC the motor could look and sound nearly stock. Darren also recommended keeping the motor below 2400cc's when building a heavy NOS or turbo motor for more cylinder wall strength. Can you imagine a 2.4-liter turbo H22 installed in a small Civic with a Hasport kit? It would make killer, non-laggy power with even a pretty low boost level. The whole thing is a pretty sick proposition.

An Armchair Analysis

With a little bit of armchair engineering analysis, I think I have figured out why this engine works so well. Because this engine has a very long stroke and a low stroke-to-rod length ratio, volumetric efficiency and power should fall off at high rpm with this motor due to the ports' limited flow capacity. However it does not. The powerband is broad and flat and the downward power trend does not start until over 7800 rpm, an unheard of rev limit for this sort of possible bore/stroke/rod length combo. We

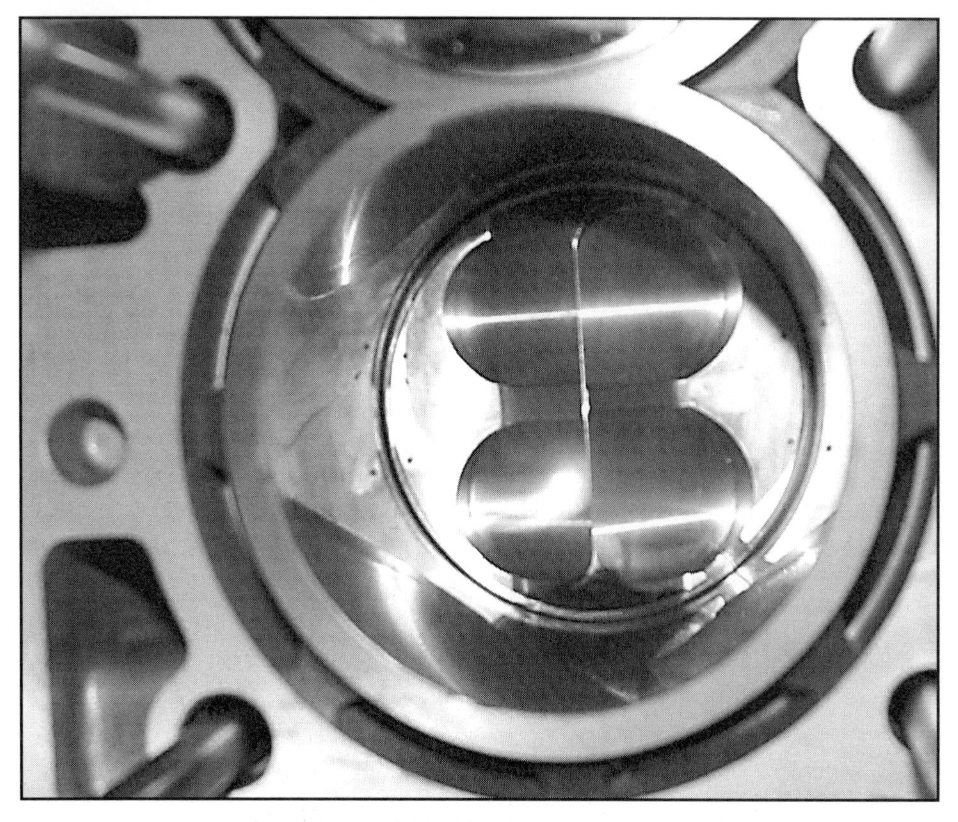

R&D uses an ultra fine finish for the bores. This helps the thin, low-tension rings seat quickly.

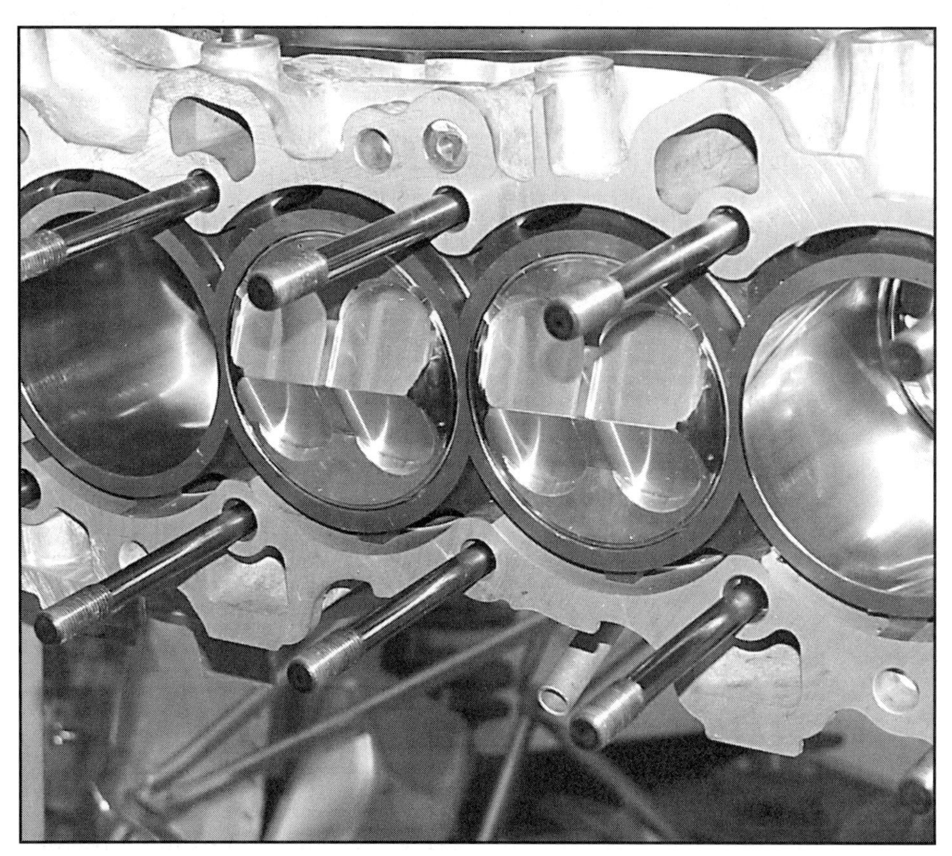

This view of R&D's B18C block shows the thick buttressed sleeves, high dome piston and studded block.

also know that combustion stability is hard to maintain at the ultra-high cylinder pressures that these astronomical compression ratios can produce.

A Matter of Volumetric Efficiency—A little known fact of engine dynamics is that volumetric efficiency increases with gains in compression ratio to a small but predictable extent. This engine has an ultra-high compression ratio and a piston speed that should not tolerate high engine rpm. As the piston speed goes up as the engine revs, the cylinder pressure starts to drop due to the falling of VE. This is due to the limitations of the cylinder heads' ports and bottom-end configuration.

When this downward trend in VE occurs, two things must happen. One, the high compression helps keeps VE reasonable, and two, the high piston speed and the dropping of VE keeps the cylinder pressures to the point where combustion stability remains manageable with super high compression. Because of the super high compression ratio, the pressure remains high enough for good thermal efficiency and power production even with the falling VE. Is R&D onto something here? Could they have found a sweet spot in the engine's dynamics, juggling VE and cylinder pressure? Darren is not talking.

Ultra high compression ratios are one of the ways that NASCAR motor builders have been able to find power in their super speedway restrictor plate motors. We think that the R&D motors are restricted by the piston speeds and ports that were not designed for the displacement or the flow demands of high piston speeds in a very similar way. Perhaps they were studying what's going on in

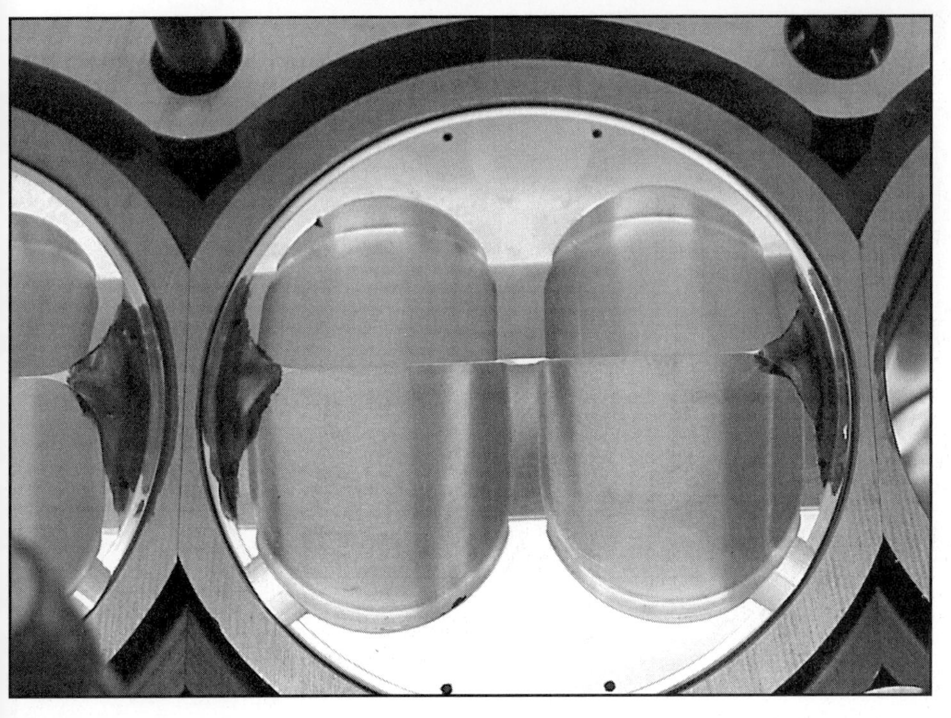

R&D gas ports the piston to help the rings maintain their seal at high piston speeds.

The 2000cc D16 motor has a full beefy girdle stock.

NASCAR and other intake restricted forms of racing??

Either way, by brute force or by really clever engineering, R&D's engines are sweet. Flat torque curves and broad useable powerbands are the signatures of R&D's monster motors. Results are what count in the end, right?

B Motors

R&D's big B motor creations can be enlarged to over 2400cc! That is a whooping 600cc larger than stock. The modification strategy is very similar to that used on the H22 with the exception of using buttressed cylinder sleeves to support the B-motors flex-prone, open-deck construction. The B-series motors run a slightly smaller lift cam and rev higher, to about 9200 rpm.

The B motors produce over 250 hp and 190 lbs-ft. of torque in race trim and a ground pounding 215 hp and 170 lbs-ft. of torque with a pump gas–friendly compression ratio of 11:1. Darren recommends a displacement limit of 2200cc's for turbo and nitrous-fired motors to maintain better general structural integrity for these more highly stressed applications.

This is all good when you consider that a B motor is usually a better-balanced package for a Civic. It is smaller and weighs much less than the big H motor and all power accessories, like air conditioning, can be run. Most of us would agree that A/C is a very nice thing to have in a street car. Even with a sano kit like Hasport's, you must usually give up AC if you want to drop a big H into a Civic.

D Motors

R&D's D16 SOHC creation that drives Bisi Ezerioha's world's fastest D motor car is built with similar modifications. R&D is getting phenomenal power out of this sometimes ignored and mostly unloved motor. This is one D motor that is not overshadowed by its bigger H- and B-series VTEC brothers.

Bisi's crank is another one of R&D's 4340 forged blanks stroked to a secret amount. The D motor uses

This inner view shows the block notching needed to clear the rods on this stroker D16.

R&D uses thick, buttressed sleeves to beef up the D16's flexy stock cylinder bores.

R&D's lightweight rods to minimize reciprocating weight. Again forged, gas ported pistons are used of an unspecified bore diameter. The engine's exact displacement is secret but we know it displaces over 2000cc's.

The block also features millwork to clear the stroker crank and ARP fasteners. Again buttressed ductile iron sleeves reinforce the Honda's weak cylinder walls and allow a big bore.

The small-port D16 cylinder head receives extensive port and combustion chamber work, while also getting 1mm larger stainless steel valves. Dual valve springs are used with Ti retainers. A special R&D cam of 0.450" lift and 312 degrees duration is used. As in R&D's B-series motors, this incredible little motor is squeezing the mixture to the tune of a 15:1 compression ratio.

In this age of high-tech programmable stand-alone fuel injection systems, Bisi's car uses a technological throwback for fueling and spark management. A good old mechanical advance distributor controls the ignition curve while a pair of two-barrel Weber DCOE carburetors handles fueling! Many of you readers probably don't even know what these are, but 10 short years ago, these were the premiere way to fuel your race car. The sight of them brings sentimental tears to my eyes! Well, Webers are still no disadvantage in power when it comes to wide open throttle operation and Bisi's car proves it. A Speed Pro EFI system is planned for the near future with twin two-barrel throttle body injectors though.

The D motor redlines at 8800 rpm and cranks out an amazing 215 hp and 170 lbs-ft. of torque to the wheels! A detuned sleeper street version of this motor can crank 170 hp and 155 lbs-ft. of torque; this figure gives respect to the little SOHC base equipment Civic motor that usually gets no respect. One of these D motors could rack up the bucks at the street races, taking the money from unsuspecting B motor pilots. Is anyone paying attention here? For turbo and NOS versions of this motor, Darren recommends keeping the displacement down to 1900cc for more strength in the cylinder walls.

THE AEBS CRX

In direct contrast to the mellow hammers of R&D are the screaming monsters built by the rebel genius of AEBS, Paulus Lee. As AEBS's chief engineer, Paulus is a true mad scientist; he speaks in staccato, his ever-working brain obviously thinking faster than his mouth can speak. He is animated and enthusiastic in his gestures. His thinking is usually several levels deeper that most people I have met in the racing industry.

The AEBS CRX driven by Roger Sangco to the championship in NIRA's All-Motor Pro-Stock Class reflects upon the unique personality of its creator. Unlike the straightforward, "no replacement for displacement" approach of R&D, the AEBS car bristles with innovation, the application of science and state of the art fresh thinking from other advanced racing venues like F-1 and CART. A deep thinker, Paulus also addresses and solves many of the problems associated with very high-revving Honda motors. There is no doubt that Paulus's theories work and that there is method behind the madness. The CRX is fast with a best pass of 10.95 @ 125 mph in the heavier NIRA Pro-Stock trim. NIRA Pro-Stock cars must run a minimum weight of 1700 lbs vs. the IDRC and IDRA minimum of 1600 lbs.

The AEBS B18C is a beautiful sight even when removed from the car.

A bottom view of the AEBS motor reveals the studded main caps and the gold color lightweight titanium rods.

The Bottom End

Let's take a look inside the AEBS motor and explain some of its finer points. Starting with the bottom end, there are plenty of tricks here; the most obvious is the AEBS block. The block uses AEBS's patented spun cast nodular iron sleeves. Spin casting gives a denser, more homogeneous metallic structure. Spin casting is where the mold is spun at a high speed so the pure dense metal is towards the outside and the slag and impurity-filled lighter metal ends up being on the inside where it is

The view from the top of the top of the block shows the thick, well braced AEBS sleeves and the AEBS studs that hold the head down.

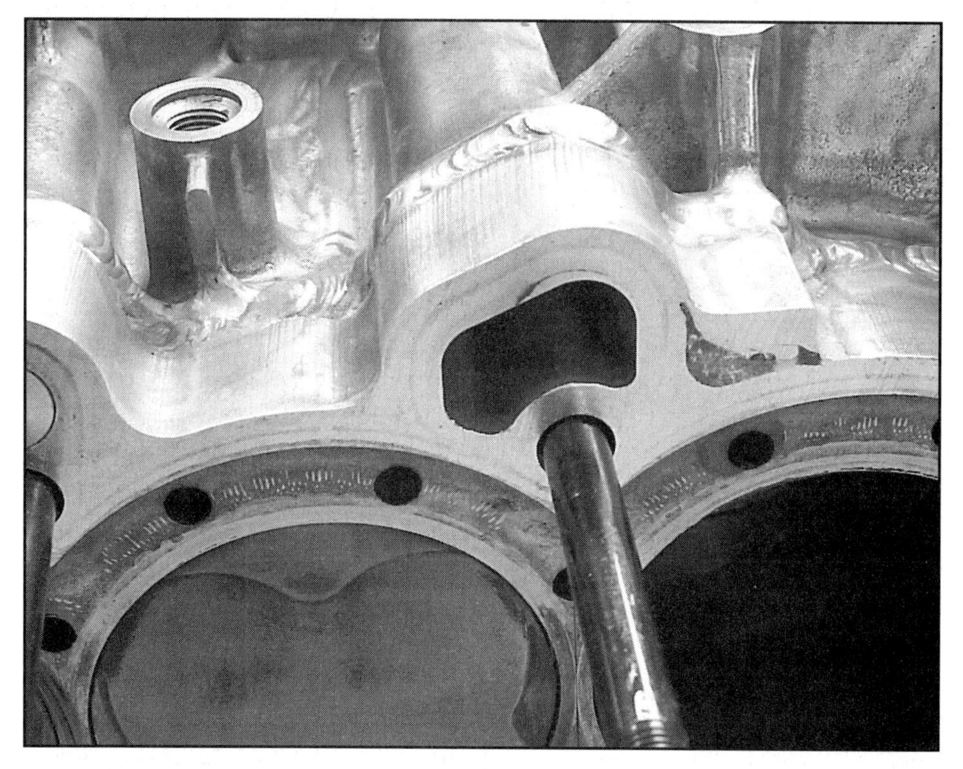

This is a close-up of the AEBS deckplate. By raising the deck, the AEBS crew was able to stroke the motor but still maintain a good stroke to rod length ratio of about 1.7:1.

removed during the final bore to size. Nodular iron is better for cylinder liners because of the large amount of carbon granules trapped within the metallic structure. This gives the liner better abrasion resistance and lubricity.

The liners also feature a very beefy, flex-free deck surface, which would make them ideal for a turbo or NOS motor. The thick deck butts solidly against the outer edges of the block, which gives the sleeves a very solid base of support to keep them from shifting under side load.

Digging a little deeper, we see some real innovation here; a thick aluminum plate, TIG-welded to the top of the block, extending the deck surface upward an unspecified amount to allow for the use of a very long connecting rod. This is a unique AEBS innovation.

The motor has a bore and stroke of 84x95 for a displacement of 2100cc's. The exact length of the connecting rod is secret but we were told that it is longer than 165mm. That would give us a stroke-to-rod length ratio of over 1.7 to 1, excellent for a Honda. The higher the rod ratio, the slower the piston dwells around TDC, helping VE at high RPM as well as aiding combustion efficiency and reducing mechanical stress on the motor. This is all pretty cool stuff, very innovative and forward thinking. This is the exact opposite of the ultra-high compression way that R&D gets around high rpm VE issues.

Crank—Looking into the bottom end we found a billet 4340 fully counterweighted AEBS stroker crank. The counterweights weights are not knife edged, just bullnosed with care being taken for the counterweights being the correct mass as an engine of this stroke may, as many Hondas that are highly stroked can have trouble with the oil pump breaking due to crank harmonics. This is amplified by the fact that many top NA motor builders cut down the crank

The AEBS-Ross pistons are hand massaged on the dome for valve and compression clearance, then treated to a thermal barrier coating to the dome. The skirts are Teflon coated for less friction and longer life.

This aluminum plate blocks off the stock water pump cavity and adapts an external electric water pump for a savings of 6 hp.

counterweights, knife edging them to lighten them and reduce windage losses. Note that both the AEBS motor and the R&D motors have fully counterweighted cranks. To further control crank torsional harmonics the harmonic balancer off of a B18B motor is used, which is the heaviest, most effective balancer.

Rods—Ultra-trick, ultra-light Cunningham titanium connecting rods hold the AEBS designed, Ross low-silicone pistons to the crank's throws. The rods feature 220,000 psi bolts and are super light in weight. These were selected to combat harmful harmonics from enhancing bending and twisting loads on the crank. Lighter rods also mean lighter counterweights are possible, a way to safely reduce power sucking reciprocating mass.

Pistons—The Ross pistons have H11 tool steel pins and Total Seal rings. The pistons are hand massaged to a high but conventional 13.5:1 compression ratio and are coated with Polymer Dynamics Teflon coat on the skirts and thermal barrier coating on the domes.

Since AEBS controls and optimizes piston speed, the piston dwells near TDC longer and the engine's displacement is smaller, a more conventional compression ratio must be used to avoid combustion stability problems. Like I said before, a diametrically opposite approach to R&D's way of doing things.

To help improve head gasket seal and to strengthen the bottom end, all of the block's structural fasteners have been replaced by an AEBS stud kit featuring 180,000 psi tensile strength studs. The block also has threaded oil passage plugs, a deburred interior and the usual signs of good engine prep. Stock Honda bearings are used for the crank and rods. The unique innovation of a decent rod ratio, harmonics control and light reciprocating weight without sacrificing strength, allow the AEBS motor to produce power to a screaming 10,000 rpm.

The block has a block-off plate and adaptors to run a trick AEBS full flow electric water pump which moves water at a 20 gallon per minute rate. The pump frees up about 6 horsepower by eliminating parasitic loses present in a crank-driven water

The cloverleaf pattern with its enlarged quench area is easily seen in this close-up of the AEBS combustion chamber. Note the gold color thermal barrier coating. Also look carefully at the valve job. You can see how the valve seats high on the unusually wide 45-degree seat cut. Venturi effect?

An overall view of the combustion chambers shows some interesting things. Note how clean and free of carbon they are after an entire year of racing. This shows that the chambers are producing an efficient, turbulent burn.

pump. A stock Type–R Integra oil pump was used for its additional lubricant flowing capacity.

The Cylinder Head

Looking at the all-important cylinder head, we find more of AEBS's engineering innovation. The combustion chambers are welded up and reconfigured so they are in a cloverleaf pattern instead of a pentroof. This was done to increase quench and to create turbulence in the combustion chamber for a more homogeneous, stabilized burn. By taking away areas for end gas to live, the engine's propensity to detonate is raised, as end gas areas at the periphery of the combustion chamber are where detonation usually starts. The high turbulence also improves combustion efficiency.

Paulus checked to make sure that the welded areas of the chamber did not cause valve shrouding by interfering with the flow around the valves on a flow bench and hand massaged the piston domes to clear the pistons' quench areas. The pistons exhaust ports and valve faces are coated with Polymer Dynamics thermal barrier coating to prevent the loss of heat from the combustion chamber.

Porting—The porting of the head is pretty straightforward with good clean ports. Nothing too radical here, just good basic headwork. Ferrera stainless steel 1 mm oversize valves live in the head featuring swirl polished heads and turned down flow improving stems. Ferrera Titanium valve spring retainers and dual valve springs control valve motion.

Valve Job—We noticed some unusual aspects of the valve job. The 45 degree seat seems slightly on the

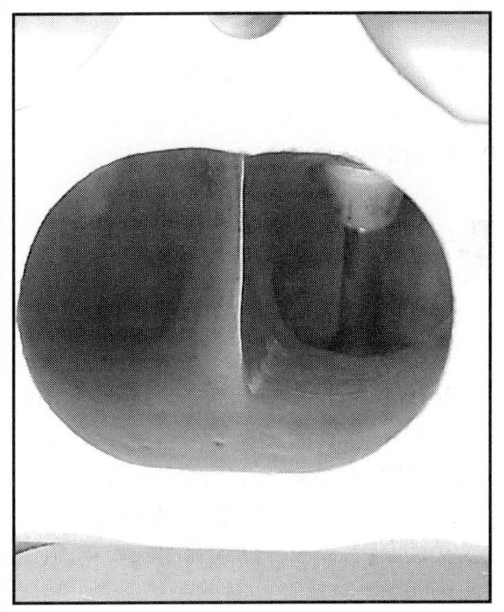

The AEBS exhaust port shows clean but conventional workmanship.

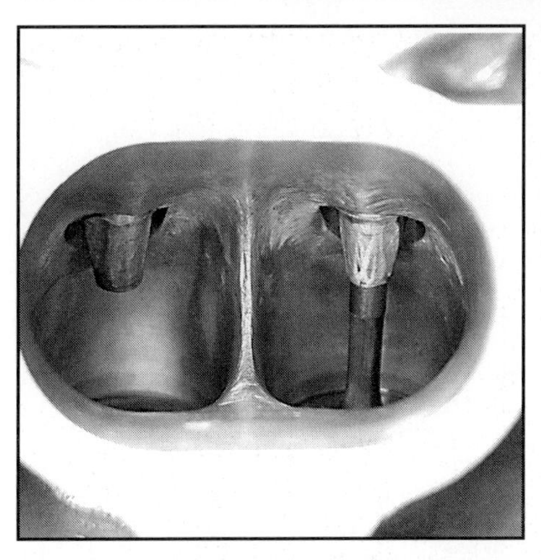

The conventionally ported intake port has bronze valve guides. Also note that the stock injector boss has been welded up flush as the injectors now preside in the intake manifold.

Dual valve springs and titanium retainers keep the 1mm oversize stainless steel valves under control.

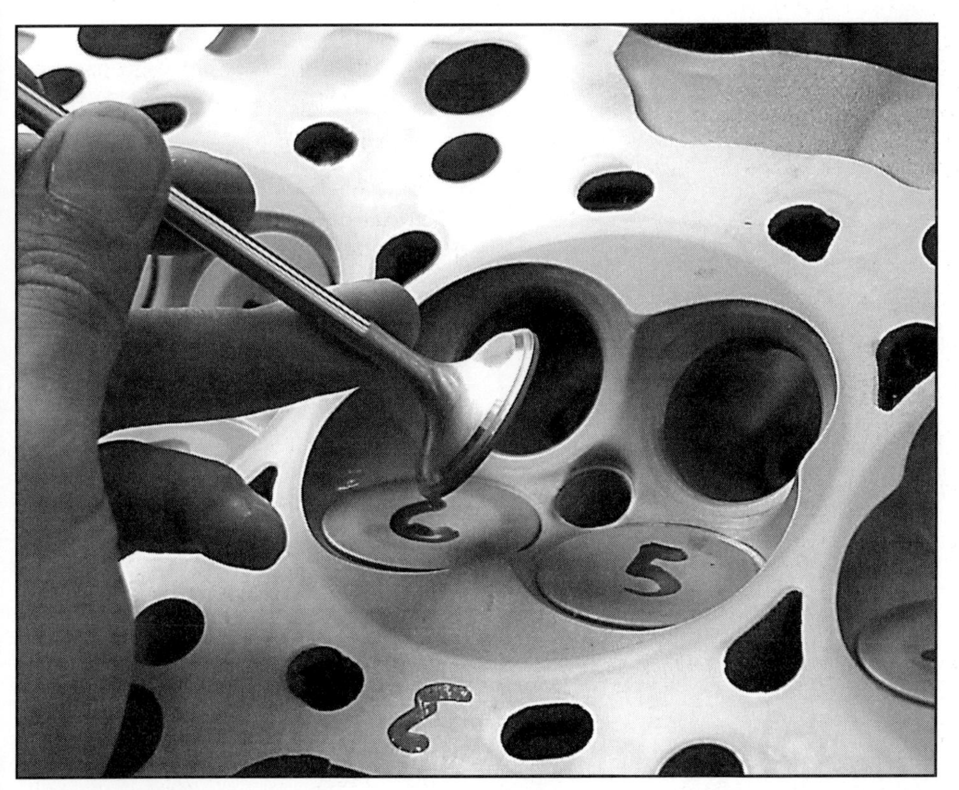

A close-up of the Stainless Ferrera valves shows the flow improving swirl polishing and the reduced diameter stem. This improves flow by blocking less of the port. The flattened profile also improves flow.

wide side and witness marks from the hand lapping jobs shows that the valve contacts high on the 45. Normally the 45 would be narrowed by the 70-degree cut right to the edge of the seating surface, but for some mysterious reason Paulus has left the 45 on the wide side towards the port side of the cut. Perhaps this was done to create a venturi effect at partial valve openings? No one is talking.

Camshaft

An AEBS camshaft moves the valves to the tune of about 345 degrees of duration and around .500" of lift. AEBS is a little secretive of the exact numbers here. Since the duration numbers sound pretty large, we figure that this is SAE duration. This would put the 0.050" lift duration at a still mind boggling 320 degrees or so. No wonder it revs to 10,000 rpm. Paulus was vague at where he sets the cams but he mumbled something like, the tightest lobe center possible. We also noticed that the two low-rpm VTEC lobes were radically different in size and shape, probably to impart some swirl on the mixture at low rpm. We know that the stock low rpm lobes are like this but on the AEBS cam, they are radically so.

Intake Manifold

The intake manifold bristles with innovation. The large hand fabricated plenum has non-parallel wall to avoid resonance. The short straight intake runners have velocity stacks that extend inside the plenum to help reduce the side effects of reversion at high rpm. The manifold features two sets of injectors. The 300cc min RC

A side view of the cam lobes reveals the marked difference between a stock Type R cam and the AEBS cam. Note the much bigger high rpm lobe on the AEBS cam with much more lift and duration than the stock lobe.

Looking down through one of the big B&K throttle bodies, you can see the velocity stacks built into the manifold as well as the secondary injector that sprays directly into the bell mouth of the stack to improve charge cooling through vaporization of the fuel.

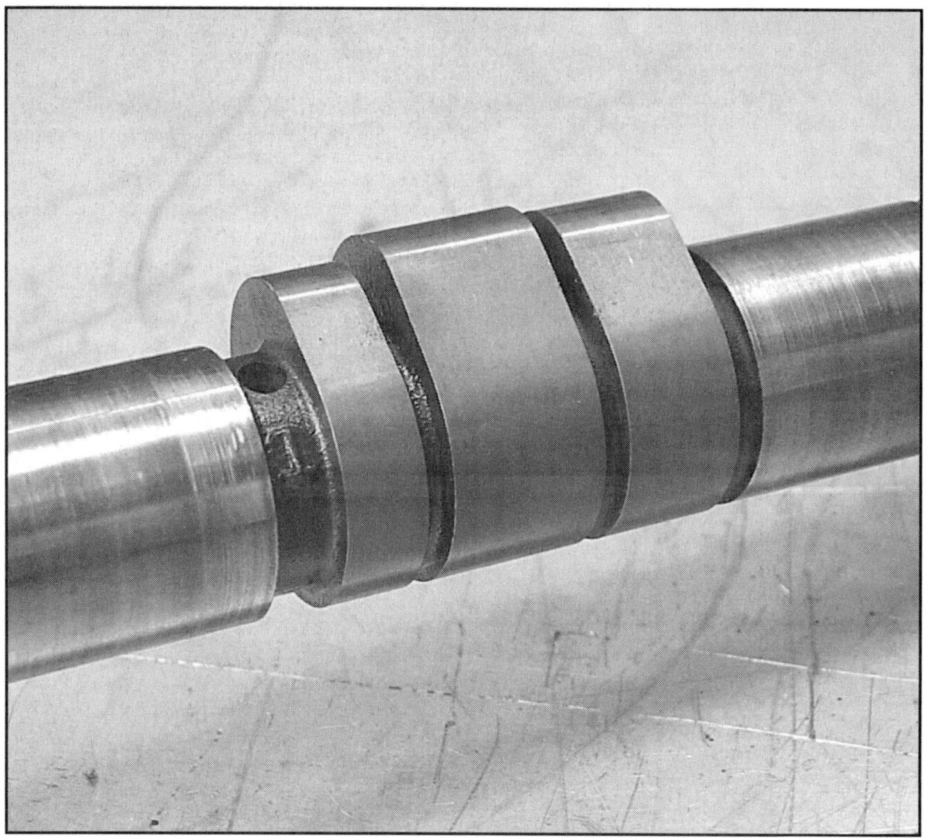

A side profile of the cam shows the differences in lift and duration in the low speed lobes. This is done to improve swirl to attain better combustion at low rpm.

engineering primary injectors are mounted on the bottom of the intake runners. This is done to place them out of the area of highest flow activity, which is usually the top of a port or intake runner. By placing the injector at the bottom, the injector tip causes less turbulence and thus better flow. Atomization is improved also. These primary injectors operate all the time.

Injectors—A second set of 300cc min injectors spray from the back wall of the plenum, directly at the bells of the velocity stacks. Paulus tells us that these injectors operate at high rpm to provide additional fuel. Their placement is to provide additional evaporative cooling much like what is done with Indy cars. Since the primary injectors are used to ensure good throttle response, the secondary injectors can be placed way back like this with no detrimental effects on throttle response.

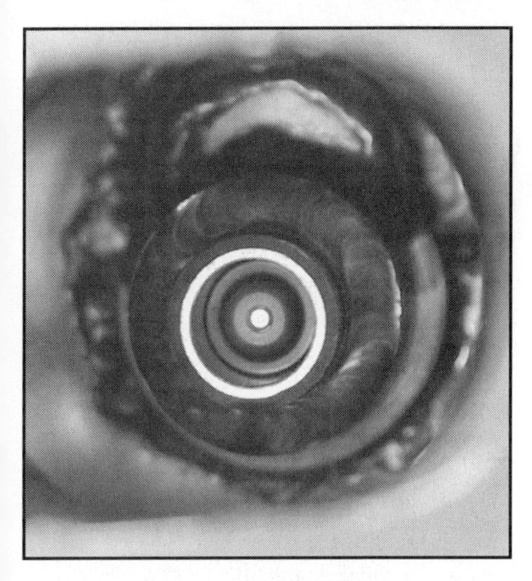

The head of the secondary injector is visible when looking up the intake port.

The thing that draws you eyes to the engine's manifold is the twin B&K throttle bodies. Each of them has sufficient flow for a V-8 engine.

The injectors are batch-fired by a SDS stand-alone ECU with the difficult switching part of the program between primary and secondary injectors, one of Paulus's secrets. Dual BK throttle bodies grace the top of the manifold. Although they are too big for this engine when calculated, their placement helps ensure equal air distribution through the manifold and helps keep the plenum non-resonate, to one degree of freedom. Finally, to help reduce heat soak, the manifold is given a coat of ceramic thermal barrier coating from Polymer Dynamics.

Exhaust

The exhaust features an AEBS header. Paulus got his start making headers and this one bristles with ideas beyond the status quo. The header is constructed of 321 stainless with the primaries stepping through 3 increasing diameters. This is to help keep the exhaust gas velocity high and to help reduce reversion on overlap. The header terminates into one of AEBS's signature collectors. Cylinders 1 and 4 are paired with cylinders 2 and 3, kept separate within

The primary injectors have been located to the bottom of the manifold's runners where they are out of the way of the bulk of the runner's flow. Here the injectors protruding tips interfere less with the flow of the ports than if they were on the roof of the runners.

A view of the massive dual intake system as installed in the car. Keep your pets and small children away from this. The large throttle area is not required for flow but is used to make sure that the distribution of air in the manifold is consistent and the plenum is kept non-resonate at high rpm.

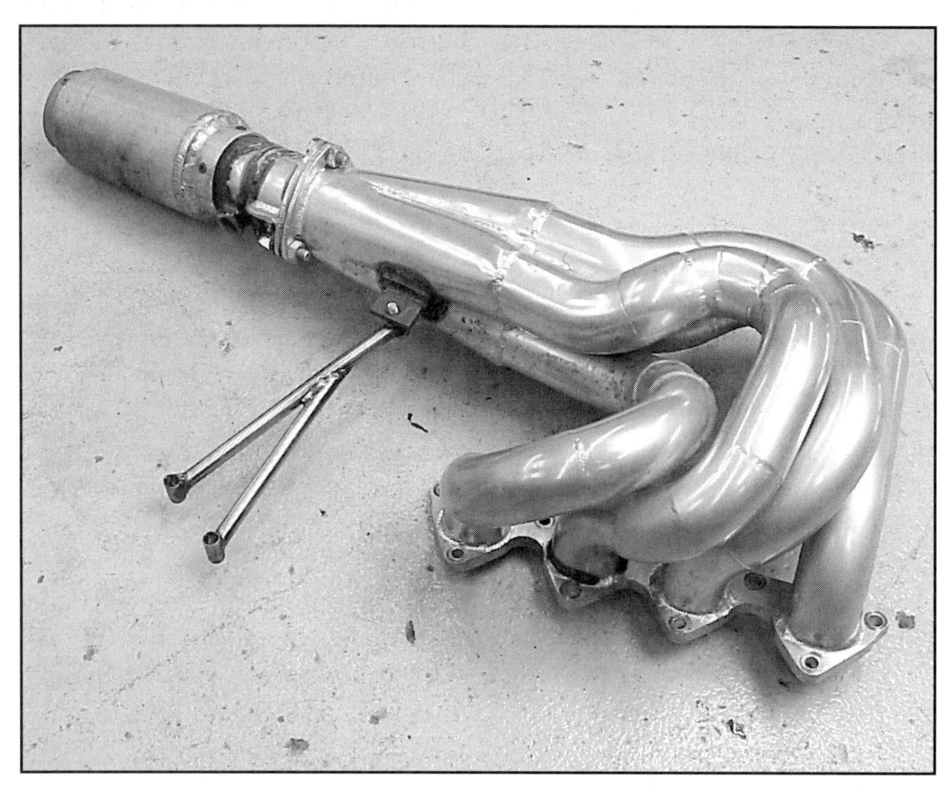

The AEBS header is a work of art with stepped diameter primaries to combat reversion as well as a secret collector that improves low-end torque with no sacrifice up top.

part of the collector by a large fin. The length of the fin is critical and secret. This provides a significant increase in torque with no other penalties.

Even the valve cover on this engine is unique. There are large diameter vents on top of the valve cover, which according to Paulus are there to allow air circulation within the motor to reduce high rpm pumping losses and to help maintain high rpm ring seal.

By the Numbers

What are the magic numbers? The wonder motor pumps out about 260 hp to the wheels at 9000 rpm with the powerband almost flat across to 10,000 rpm. The motor cranks out 166 lbs-ft. of torque, peaking at about 7200 rpm.

Predictably, the AEBS motor has a slight edge in the horsepower department, by about 2–5 peak hp over the R&D motor. However the R&D motor puts out almost 20 lbs-ft. more torque over a much broader powerband. Its the area under the curve more than peak hp that wins races. Other than the fuel requirements, the R&D motor's powerband would make it a pleasant motor to kick some ass with on the street! In direct contrast the AEBS screamer does not come alive until 6500 rpm.

The folks at AEBS were proud to point out that the motor ran the entire 2000 season without a single failure. The valve cover never even came off. The only things that needed to be checked were the plugs. This is not like many of the other all-motor competitors who were plagued by failure during the year. Paulus pointed out that the engine was still in excellent condition, with good compression and leak-down numbers.

The engine was just pulled to freshen it up, as it will be sold to another competitor to be replaced with one of his even more radical creations.

COMPARING THE TWO

The motors from R&D and AEBS are diametrically opposed in their engineering and characteristics. You are probably wondering what it all boils down to, which one is faster. Surprisingly, it is almost a toss-up. In practice, Jeremy's car is the faster of the two but Jeremy runs mostly in IDRC and IDRA, where his car can be 100-lbs. lighter. Roger runs solely in NIRA and must carry 1700 lbs. If both cars weighed the same, it would be very interesting indeed. The true answer may never be known as Roger's car and motor have been sold by AEBS and AEBS is concentrating on other secret programs, although they will still build customer motors to this spec.

The torque of the R&D car would be less affected by ballast we think and its ultra-high compression motor would fair better at high altitudes. Its broad powerband may be more forgiving to driver error while traveling down the 1320. However the lower torque of the AEBS car may make it easier on the driver to launch. In either case, the race would be a close one.

The R&D motor would be a superior motor in street trim. However, if the compression was reduced to around 11:1, the AEBS motor would be the one wailing 10,000 rpm street machine, still having more low-end torque than a stock B18 motor due to its larger displacement.

THE CRYSTAL BALL

Let us look to the future and make some predictions of where naturally aspirated racing may go in the next few years. In this young and innovative class, the formula for a winning car is not set yet as the two cars we have just studied shows. There will be much study and much more development before a winning formula has stabilized, like it has in the turbocharged classes.

We predict that more attention will be made to the engine's stroke-to-rod length ratio and bore-to-stroke ratio in an attempt to reduce piston speed in Hondas. Bigger displacement, higher-revving engines will result.

Careful study of the control valvetrain harmonics and an increased use of lightweight titanium will allow higher revs, perhaps in the 11–12,000 rpm range. A major domestic aftermarket company will come out with an improved cylinder head for the workhorse Honda B-series motor with high velocity, good flowing ports and a low surface area to chamber volume combustion chamber with turbulent combustion inducing quench pads.

Roller cam followers will debut and be widely used. This will free up horsepower by reducing valvetrain friction. The roller itself will allow a more radical cam profile for more area under the cam's lift curve. These sorts of cams will allow a broader, higher power curve.

More attention will be given to crankcase pumping losses. In Formula 1 racing, much research has been applied in this area with significant gains as a result. We predict that a partitioned crankcase divided between cylinders 2 and 3 will help considerably at high rpm.

Dry sump oil systems from road racing will increase power by 5–10 hp due to better ring seal and less windage losses. Dry sumps will be needed to ensure good oiling at over 10,000 rpm anyway. If no one invests in dry sumps, a stronger oil pump is needed, as many of the higher revving NA Hondas are breaking oil pump gears due to crank harmonics.

Special gasoline-based fuels will help stabilize combustion and speed the flame-front travel without detonation. European F-1 fuel supplier Elf is reportedly considering bringing some of their super fuels to the US market.

Sequentially shifted, straight-cut-gear, constant mesh transaxles with selected ratios will help engine builders by allowing them to build more powerful, higher revving engines without worrying about gear spacing. Straight-cut gears will soak up less power in the transaxle. The issue is who will be able to afford this technology.

Incremental gains will take place in exhaust design, reciprocating weight reduction and suspension tuning. With these developments, we should be seeing low 10s pretty soon; perhaps times will be knocking on the 9s if the constant mesh transmission is embraced.

Who knows if any of these predictions will come true? But one thing is for sure, the performance envelope of the Honda/Acura engine will continue to be pushed far beyond anything we've seen yet!

HONDA PERFORMANCE DIRECTORY

Note: *What follows is just a partial list of the many aftermarket suppliers that offer parts and accessories for the Honda/Acura enthusiast. Check your favorite magazines such as* Sport Compact Car, Super Street *and* Import Tuner *for a more complete listing of manufacturers and suppliers.*

Advanced Engine Management (AEM)
200 Corporate Pointe, Ste. 465
Culver City, CA 90230
Phone/Fax: (310) 327-9336/258-0036
Internet: www.a-e-m.com
AEM is best known for their Tru-Time adjustable cam timing gears, Tru-Power ancillary component pulleys, and cold air intake system. AEM also has its own bottom end engine kits, including pistons, rods and engine stud kits for racing applications. AEM also owns CoolTech, a leading brake systems manufacturer. The company now has a kit for Hondas featuring 12.6-inch rotors and 4-piston calipers.

AEBS
8270 Miramar Road
San Diego, CA 92126
Phone: (858) 693-3200
Internet: www.aebsys@aol.com
AEBS is an innovator in maximum Honda power production. AEBS has built some of the fastest championship winning Honda motors in the country. AEBS is a distributor of Hondata user programmable ECU's, Tough Ross forged pistons and Tomei products.

A'PEXi
130 McCormick Avenue, Ste. 107
Costa Mesa, CA 92626
Phone/Fax: (714) 444-4378/444-4379
Internet: www.apexi-usa.com
Offers a complete line of products including cat-back exhaust systems, turbo systems, boost controllers, fuel controllers, ignition timing computers, air filters, complete coil-over suspensions, and aerodynamic body kits.

Arias Forged Racing Pistons
13420 S. Normandie Avenue
Gardena, CA, 90249
Phone/Fax: (310) 532-9737/516-8203
Internet: www.ariaspistons.com
Arias makes high quality forged pistons that are used by many successful Honda racers.

Burns Stainless
1013 W. 18th St.
Costa Mesa, CA 92627
Phone/Fax: (949) 631-5120/631-3184
Internet: www.burnsstainless.com
A supplier of stainless steel fabrication bits, including U and other mandrel bends, Burns also has pre-made merged collectors and more exotic tubing like titanium and inconel.

Carrillo Industries
990 Calle Amanecer
San Clemente, CA 92673
Phone/Fax: (949) 498-1800/498-2355
Internet: www.carrilloind.com
Carrillo makes the finest connecting rods available. Used from Indy to Bonneville you will be hard pressed to find a stronger, more durable connecting rod.

Comptech
4717 Golden Foothill Parkway
El Dorado Hills, CA 95762
Phone/Fax: (916) 939-9118/939-9196
Internet: www.comptechusa.com
Features Comptech springs, shocks and anti-roll bars. Also a big brake kit for Civics, Accords and Preludes featuring Brembo calipers and big rotors.

Crane Cams
530 Fentress Blvd.,
Daytona Beach, FL 32114
Phone/Fax: (386) 252-1151/947-5554
Internet: www.cranecams.com
Crane is best known for their camshafts and valvetrain parts. Crane makes one of the only billet cams for the SOCH D series motor. Crane also makes and excellent high-powered ignition system.

Crower
3333 Main Street
Chula Vista CA 91911
Phone/Fax: (619) 422-1191/422-9067
Internet: www.crower.com
Crower makes a wide assortment of billet rods, cranks, camshafts and all assorted engine internal parts. Most of the leading Honda powered racers use some sort of Crower Internals.

Cunningham Rods
550 W. 172nd Street
Gardena, CA 90248
Phone/Fax: (310) 538-0605/538-0695
Internet: www.cunninghamrods.com
Cunningham rods are available in super light titanium as well as billet chrome moly. Cunningham makes an extremely high quality rod. Ask Lisa Kubo.

Darton International
2380 Camino Vida Roble, Ste. J&K
Carlsbad, CA 92009
Phone/Fax: (760) 603-9895/603-9629
Internet: www.darton-international.com
Darton makes bulletproof cylinder sleeves with an integrated block guard , which helps make near bulletproof bottom ends.

DC Sports
286 Windfield Circle
Corona, CA 91720
Phone/Fax: (909) 734-2030/734-2792
Internet: www.dcsports.com

Best known for their killer stainless headers and exhaust systems, DC Sports also makes timing gears, fuel rails and suspension parts.

DRP Racing Heads
1015 1/2 West 190th St.
Gardena CA, 90248
Phone/Fax: (310) 523-4074/833-3727
Internet: www.dpr-racing.com
Dan Paramore has been porting heads since time began. Dan has been buds with the author since they worked mainly on bicycles. DPR heads have been found on quite a few fast Hondas and other race and streetcars across the country. Dan was instrumental in helping with this book by providing technical information, pictures and whatever was needed to be scrounged up in short notice.

Dynojet
2191 Mendenhall Dr.
North Las Vegas, NV 89031
Phone/Fax: (800) 711-4498/(702) 633-4821
Internet: www.dynojet.com
Dynojet makes one of the most repeatable chassis dynos that can handle high-powered cars. You can look up Dynojet on the web or call to find the location of the closest Dynojet-equipped shop near you.

Edelbrock
2700 California Street
Torrance, CA 90503
Phone/Fax: (310) 781-2222/320-1187
Internet: www.edelbrock.com
Edelbrock, a mainstay of domestic performance, has teamed up with JG, pioneers of Honda performance to develop a line of parts for Honda motors. Edelbrock has headers, exhaust systems, manifolds, nitrous systems and many other parts for Hondas.

Eshleman's Carbon Works
42207 6th West, Unit 1012
Lancaster, CA 93534
Phone/Fax: (661) 948-5155/948-7910
Internet: www.thecarbonworks.com
Eshleman's makes a killer, super-light, carbon fiber, short-runner intake manifold for Hondas, as well as valve covers.

Extrude Hone®
8800 E. Somerset Blvd
Paramount, CA 90723
Phone/Fax (562) 531-2976/531-8403
The Extrude Hone® abrasive porting method is one of the secrets behind Lisa Kubo's domination of the Quick Class. Extrude honing cannot be surpassed in the porting of intake manifolds and turbocharger components.

Ferrea Racing Components
2600 N.W. 55 Court, Ste. 238
Ft. Lauderdale, FL 33309
Phone/Fax (954) 733-2505/735-2179
Internet: www.ferrea.com
Ferrea makes valvetrain components, namely valves, valvesprings and titanium retainers, for most Honda motors.

Fluidyne
4850 East Airport Drive
Ontario, CA 91761
Phone/Fax (800) Fluidyne/(909) 390-3950
Internet: www.fluidyne.com
Fluidyne makes trick high capacity racing aluminum radiators that bolt right into most Honda applications.

F-Max Fabrication
910 Stanley Ave.
Escondido, CA 92026
Phone/Fax (760) 746-6638/746-6682
Internet: www. f-max.com
F-max is a premier fabricator that can do one off custom work. They also have high quality and complete turbo kits available for most Honda applications.

Garrett Engine Boosting Systems
3201 W. Lomita Blvd.
Mail Stop U5
Torrance, CA 90505-5064
Garrett manufactures some of the best turbochargers in the world. Don't bother trying to call, they don't sell to the public.

GReddy
9 Vanderbilt
Irvine, CA 92618
Phone/Fax: (949) 588-8399/588-6318
Internet: www.greddy.com
GReddy's main product line is centered on cat-back exhaust systems with stainless steel cans and mild steel tubes. They also have a header system for Honda and Acura cars with CARB approval. GReddy has a street legal turbocharger kit for '92–'95 Honda Civic 1.6-liter cars featuring 5.5-psi boost. Other products the company sells include oil cooler kits, strut tower bars, air filters, interior accessories, 20-30 mm drop chassis springs.

Hasport Performance
3030 S. 40th St. Unit #1
Phoenix, AZ 85040
Phone/Fax (602) 470-0065/470-0516
Hasport makes absolutely the best kits to do hybrid motor swaps, putting big motors in small Civics. Hasport also has connections to supply used Integra B series and Prelude H series motors to help you make your swap. Hasport also has a line on genuine OEM JDM Honda parts.

Hi-Tech Exhaust
12 Hammond Drive, Ste. 202
Irvine, CA 92618
Phone: (949) 581-2181
Internet: www.hi-techexhaust.com
Hi-Tech makes some of the world's best headers and turbo manifolds, crafted from 321 stainless. Hi-Tech is the little known power secret behind many of the world's fastest Hondas. Hi-Tech pioneered anti-reversion technology. Hi-Tech also has a new Honda B series roller cam.

HKS USA Inc.
2355 Mira Mar Ave.
Long Beach, CA 90815
Phone/Fax: (562) 494-8068/494-1768
Internet: www.hksusa.com
HKS offers a complete range of high performance products for Hondas/Acuras, including cat-back exhaust systems, Super Mega-Flow intake, adj, cam gear, and HKS Twin-Power ignition.

HP Racing
1590 N.W. 108 Ave.
Miami, FL 33172
Phone/Fax: (305) 629-8977/594-4235
Internet www.hpracing.com
HP Makes all sorts of bolt-ons as well as large plenum manifolds.

Innovative Turbo Systems
845 Easy Street, #102
Simi Valley, CA 93065
Phone/Fax: (805)526-5400/526-9240
Innovative is a distributor for Garrett turbochargers as well as an innovator, offering their own dual ball bearing center section, wastegates and boost controls. Innovative offers turbines in high temperature inconel and Mar M with housings made of Ni-Resist, stuff that most other aftermarket turbo companies don't bother with. Innovative sells some of the newer Garrett GT turbos as well as the venerable TO4/T3.

Jackson Racing
440 Rutherford Street
Goleta, CA 93117
Phone/Fax: (888) 888-4079/692-2525
Internet: www.jacksonracing.com
Jackson Racing was one of the pioneers of Honda performance, and their line of products has grown considerably to include suspension kits and exhausts. However, they have concentrated mainly on their supercharger program for Honda and Acura cars, a program that is unequaled in terms of quality, fit and finish.
Jackson's supercharger kits are based on an Eaton blower. The supercharger kits typically run 6 psi boost contributing to about a 40 percent increase in horsepower.

JE Pistons
15312 Connector Lane
Huntington Beach, CA 92649
Phone: (714) 898-9763
Internet: www.jepistons.com
JE Pistons are a popular, high quality forged piston.

JG Engine Dynamics
431 S. Raymond Ave. #102
Alhambra, CA 91803
Phone/Fax: (626) 281-5326/281-5306
JG Engine Dynamics specializes in custom machine work including heads, cam grinding, block modifications and engine building, all backed up with thorough dyno testing.

Jun USA Inc.
1398 29th Street
Signal Hill, CA 90806
Phone/Fax (562) 424-7828/424-6360
Jun is perhaps the premium supplier of high-end JDM motor parts and engine building. Jun has perhaps the best pistons, cams, cranks, rods, valves, manifolds, valve springs and other parts on the market. Although expensive, the quality of Jun components is exquisite.

King Motorsports
105 East Main Street
Sullivan, WI 53178
Phone/Fax: (262) 593-2800/593-2627
Internet: www.kingmotorsports.com
King Motorsports is the American distributor for Mugen parts. Mugen is a company founded by the son of Mr Honda! Although their parts are expensive you can be assured that they work well. King Motorsports also has full service engine building and race car prep with a long series of SCCA championships under their belt.

K&N
561 Iowa Ave
Riverside, Ca 92502
Phone/Fax (909) 684-9762/684-0716
Internet: www.knfilters.com
K&N is THE supplier of high performance, high flow air filters. Accept no substitute.

Magnaflow
22961 Arroyo Vista
Rancho Santa Margarita, CA 92688
Phone/Fax (949) 858-5900/858-3600
Internet:www.magnaflow.com
Magnaflow makes high quality, stainless mufflers as well as mandrel bends and exhaust supplies.

MoTec Systems USA
5355 Industrial Drive
Huntington Beach, CA 92649
Phone/Fax: (714) 897-6805/897-8782
Internet: www.motec.com
According to the movie, "The Fast and The Furious", MoTec makes awesome exhaust systems; nothing could be further than the truth. MoTec makes perhaps the premier, stand-alone engine management systems available to the public, not exhausts.

MSD Ignitions
Autotronic Controls Corporation
12120 Esther Lama, Ste. 114,
El Paso, Texas 79936
Phone: (915) 855-7123
Internet: www.msdignitions.com
In my opinion, MSD makes the best ignition systems available. Powerful and reliable, MSD can fire the spark through the highest boost and loads of NOS.

NOS/Holley Performance Products
1801 Russellville Road
P.O. Box 10360
Bowling Green, KY 42102-7360
Phone: (270) 782-2900
Internet: www.holley.com
The originators and always innovators of nitrous oxide injection, NOS is now part of the Holley carburetor empire.

Place Racing
1016 West Gladstone Street
Azusa, CA 91702
Phone/Fax: (626) 334-3345/334-3225
Internet: www.placeracing.com
Place Racing has a line of high-quality engine swap kits for Civics and other Hondas. Place Racing also sells heavy-duty urethane motor mounts and cold air intakes.

Port Flow Design
1583 W. 259th st
Harbor City, CA 90710
Phone/Fax: (310) 257-0351/257-0352

Internet: www.portflow.com
Port Flow Design cylinder heads are found on the quickest Hondas in the country, including the quickest All Motor class Honda.

Powerdyne
104-C East Ave., Ste. K-4
Lancaster, CA 93535
Phone/Fax: (661) 723-2800/723-2802
Internet: www.powerdyne.com
Powerdyne has a centrifugal supercharger kit available for D series Civics. The Powerdyne kit claims to add 80 hp to your Civic.

RC Engineering
1728 Border Ave.
Torrance, CA 90501
Phone/Fax: (310) 320-2277/782-1346
Internet: www.rceng.com
RC Engineering has an extensive injector cleaning, balancing service as well as a good selection of various fuel injection components.

R&D Racing
115 E. Gardena Blvd.
Gardena, CA 90248
Phone: (310) 516-1003
R&D is a builder and supplier of parts for the fastest All Motor Honda engines in the country. Their large displacement conversions make excellent and deadly street and turbo motors as well. R&D offers extensive dyno faculties as well as head porting and engine machining.

RS Motorworks
22515 S. Vermont Ave. #D
Torrance, CA 90502
Phone: (310) 328-5514
RS Motorworks is known for their block plates, allowing a stroked motor with long connecting rods for a good stroke-to-rod length ratio and low piston speeds.

Skunk 2 Racing
25710 Industrial Blvd. #2
Hayward, CA 94545-2920
Phone/Fax: (510) 781-0538/781-0539
Internet: www.skunk2racing.com
Skunk2, formally known as the Skunkworks, is the originator of the hyper-quick, All Motor cars, having the first 10-second, naturally aspirated class racer. Skunk 2 has a variety of innovative parts like a short runner manifold and billet camshafts.

SPW Industries
201 Bernoulli Circle, Bldg. D
Oxnard, CA 93030
Phone/Fax: (805) 485-5249/604-0249
www.spwusa.com
SPW stocks a wide assortment of high performance, genuine Honda JDM high performance parts for the Acura Integra B series and Prelude H series motor.

Steve Millen Sports Parts
3176 Airway Ave.
Costa Mesa, CA 92626
Phone/Fax: (714) 540-9154/540-5784
Internet: www.stillen.com
Stillen is the king of bolt-on street performance parts. If you are just starting out, Stillen is a one-stop performance shop.

STR Speedlab
949 North Cataract Ave., Unit F
San Dimas, CA 91773
Phone/Fax: (909) 394-4719/394-1739
Internet: www. strspeedlab.com
STR makes an assortment of engine parts, from cylinder sleeves to intake manifolds.

Tial Sport
510 Washington St., #16
Owosso, MI 48867
Phone/Fax: (517) 729-8553/729-9973
Internet: tialinc@earthlink.net
Tial is the maker of the best wastegates and blow-off valves in the world. Reliable and trick, these wastegates are found on some of the quickest Hondas in the world, like those of Lisa Kubo and Steff Papadakis.

Toda Racing
AKH Trading
192 Technology Drive, Ste. V
Irvine, CA 92618
Phone/Fax (949) 450-1056/450-1059
Internet: www.todaracing.com
Toda makes some of the best B-Series VTEC camshafts available as well as lots of other hard-core internal engine parts.

Turbonetics
2255 Agate Court
Simi Valley, CA 93065
Phone/Fax: (805) 581-0333/584-1913
Internet: www.turboneticsinc.com
Turbonetics turbos are used by more winning turbo-powered Hondas than any other. Turbonetics originated the ball bearing center section. Turbonetics is a Garrett distributor and has a wide range of turbo combinations on the shelf and in stock for quick ordering.

TWM Induction
325D Rutherford Street
Goleta, CA 93117
Phone/Fax: (805) 967-9478/683-6640
Internet: www.twminduction.com
TWM is a source for throttle bodies and individual runner intake manifolds.

Unorthodox Racing
45 D Nancy St.
West Babylon, NY 11704
Phone/Fax: (631) 253-4909/253-4907
Internet: www.unorthodoxracing.com
Unorthodox is the originator of import underdrive pulleys. Unorthodox is also producing a set of stronger oil pump gears for Hondas.

Venom Performance
8625 Central Ave., Ste. J
Stanton, CA 90680
Phone: (800) 959-2865
Internet: www.venom-performance.com
Venom makes a sophisticated, computer-controlled nitrous system as well as intake manifolds and other various fuel system products.

Vortech
1650 Pacific Ave.
Channel Islands, CA 93033-9901
Phone/Fax: (805) 247-0226/247-0669
Internet: www.vortechsuperchargers.com
Vortech makes high-powered centrifugal superchargers for Honda engines.

ZDYNE LLC
6931 Topanga Canyon Blvd.,Ste. 8
Canoga Park, CA 91303
Phone (818) 888-5350
Internet: www.zdyne.com
Zdyne makes a very trick, modified factory ECU that can be user-programmed with a very user-friendly interface.

Zex Nitrous Systems
3418 Democrat Road
Memphis, TN 38118
Phone: (888) 817-1008
Internet: www.zex.com
Zex makes potent and user-friendly nitrous systems.

Mike Kojima has been working hard at blowing up cars throughout his 21-year professional automotive career. He has always had a passion for things mechanical from an early age. As a small child, Mike was adept at taking expensive things apart and not always being able to put them back together, much to his parents' dismay. Some things never change, except that now the things he takes apart are much bigger and much more expensive.

Mike's professional career has been wide and varied, with stints as an engineer at Toyota Racing Development (TRD) and Nissan North America as well as at Pep Boys. Mike's writing can be found regularly in the pages of *Sport Compact Car* magazine, *Turbo, Honda Tuner, speedoptions.com, sentra.net, se-r.net, TMR* magazine and *Grass Roots Motorsports* magazine.

Mike has also held SCCA and IMSA racing licenses and has participated in just about every form of racing from offroad to autocross, but above all, has been a pioneering import performance nut since the '70s.

Mike is perhaps best known for his famous quote, "Do you want to go fast or do you want to suck?"

OTHER BOOKS BY HPBOOKS

HANDBOOKS

Auto Electrical Handbook: 0-89586-238-7 or HP1238
Auto Upholstery & Interiors: 1-55788-265-7 or HP1265
Car Builder's Handbook: 1-55788-278-9 or HP1278
Powerglide Transmission Handbook:1-55788-355-6 or HP1355
Turbo Hydramatic 350 Handbook: 0-89586-051-1 or HP1051
Welder's Handbook: 1-55788-264-9 or HP1264

BODYWORK & PAINTING

Automotive Detailing: 1-55788-288-6 or HP1288
Automotive Paint Handbook: 1-55788-291-6 or HP1291
Fiberglass & Composite Materials: 1-55788-239-8 or HP1239
Metal Fabricator's Handbook: 0-89586-870-9 or HP1870
Paint & Body Handbook: 1-55788-082-4 or HP1082
Sheet Metal Handbook: 0-89586-757-5 or HP1757

INDUCTION

Bosch Fuel Injection Systems: 1-55788-365-3 or HP1365
Holley 4150: 0-89586-047-3 or HP1047
Holley Carbs, Manifolds & F.I.: 1-55788-052-2 or HP1052
Rochester Carburetors: 0-89586-301-4 or HP1301
Turbochargers: 0-89586-135-6 or HP1135
Weber Carburetors: 0-89586-377-4 or HP1377

PERFORMANCE

Baja Bugs & Buggies: 0-89586-186-0 or HP1186
Big-Block Chevy Performance: 1-55788-216-9 or HP1216
Big-Block Mopar Performance: 1-55788-302-5 or HP1302
Bracket Racing: 1-55788-266-5 or HP1266
Brake Systems: 1-55788-281-9 or HP1281
Camaro Performance: 1-55788-057-3 or HP1057
Chassis Engineering: 1-55788-055-7 or HP1055
Chevy Trucks: 1-55788-340-8 or HP1340
Ford Windsor Small-Block Performance: 1-55788-323-8 or HP1323
Honda/Acura Engine Performance: 1-55788-384-X or HP1384
High Performance Hardware: 1-55788-304-1 or HP1304
How to Hot Rod Big-Block Chevys: 0-912656-04-2 or HP104
How to Hot Rod Small-Block Chevys: 0-912656-06-9 or HP106
How to Hot Rod Small-Block Mopar Engines: 0-89586-479-7 or HP1479
How to Hot Rod VW Engines: 0-912656-03-4 or HP103
How to Make Your Car Handle: 0-912656-46-8 or HP146
John Lingenfelter: Modify Small-Block Chevy: 1-55788-238-X or HP1238
Mustang 5.0 Projects: 1-55788-275-4 or HP1275
Mustang Performance (Engines) 1-55788-193-6 or HP1193
Mustang Performance 2 (Chassis): 1-55788-202-9 or HP1202
Mustang Performance Engine Tuning: 1-55788-387-4 or HP1387
1001 High Performance Tech Tips: 1-55788-199-5 or HP1199
Performance Ignition Systems: 1-55788-306-8 or HP1306
Race Car Engineering & Mechanics: 1-55788-064-6 or HP1064
Small-Block Chevy Performance: 1-55788-253-3 or HP1253

ENGINE REBUILDING

Engine Builder's Handbook: 1-55788-245-2 or HP1245
Rebuild Air-Cooled VW Engines: 0-89586-225-5 or HP1225
Rebuild Big-Block Chevy Engines: 0-89586-175-5 or HP1175
Rebuild Big-Block Ford Engines: 0-89586-070-8 or HP1070
Rebuild Big-Block Mopar Engines: 1-55788-190-1 or HP1190
Rebuild Ford V-8 Engines: 0-89586-036-8 or HP1036
Rebuild GenV/Gen VI Big-Block Chevy : 1-55788-357-2 or HP1357
Rebuild Small-Block Chevy Engines: 1-55788-029-8 or HP1029
Rebuild Small-Block Ford Engines: 0-912656-89-1 or HP189
Rebuild Small-Block Mopar Engines: 0-89586-128-3 or HP1128

RESTORATION, MAINTENANCE, REPAIR

Camaro Owner's Handbook ('67–'81): 1-55788-301-7 or HP1301
Camaro Restoration Handbook ('67–'81): 0-89586-375-8 or HP1375
Classic Car Restorer's Handbook: 1-55788-194-4 or HP1194
How to Maintain & Repair Your Jeep: 1-55788-371-8 or HP1371
Mustang Restoration Handbook ('64 1/2–'70): 0-89586-402-9 or HP1402
Tri-Five Chevy Owner's Handbook ('55–'57): 1-55788-285-1 or HP1285

GENERAL REFERENCE

A Fan's Guide to Circle Track Racing: 1-55788-351-3 or HP1351
Auto Math Handbook: 1-55788-020-4 or HP1020
Corvette Q&A: 1-55788-???-?
Ford Total Performance, 1962–1970: 1-55788-327-0 or HP1327
Guide to GM Muscle Cars: 1-55788-003-4 or HP1003

MARINE

Big-Block Chevy Marine Performance: 1-55788-297-5 or HP1297
Small-Block Chevy Marine Performance: 1-55788-317-3 or HP1317